Instrumental Realism

The Indiana Series in the Philosophy of Technology

Don Ihde, general editor

Instrumental Realism

The Interface between Philosophy of Science and Philosophy of Technology

Don Ihde

INDIANA UNIVERSITY PRESS
Bloomington and Indianapolis

The paper used in this publication meets the minimum requirements of American National Standard for Information Sciences—Permanence of Paper for Printed Library Materials, ANSI Z39.48-1984.

∞™

Manufactured in the United States of America

Library of Congress Cataloging-in-Publication Data

Ihde, Don, date.
Instrumental realism : the interface between philosophy of science and philosophy of technology / Don Ihde.
p. cm. — (Indiana series in the philosophy of technology)
Includes bibliographical references and index.
ISBN 0-253-32899-3. — ISBN 0-253-20626-X (pbk.)
1. Technology—Philosophy. 2. Science—Philosophy. I. Title.
II. Series.
T14.I352 1991
601—dc20

90-42334
CIP

1 2 3 4 5 95 94 93 92 91

For Gabriel

Contents

Preface

Nearing its sixth decade as a subdiscipline in philosophy, the philosophy of science clearly can be said to be established. Now at the end of its first decade as an organized subdiscipline, the philosophy of technology is still having birth pangs. This monograph, a primer really, enters an interstice between these two subspecializations with several purposes in mind:

First, not only is there a difference between middle age and youth in the two subdisciplines, but the two are only vaguely related. The philosophy of science, today becoming more diverse than it once was, nevertheless has had a strong lineage connection with the Anglo-American philosophies. Positivism—although of continental birth—became strong when it fled the continent and found itself in a favorable position in North American universities in the very beginnings of this subdiscipline. Later, other analytically inclined directions became revisionist but remained dominant. By contrast, the loose groupings of philosophers who were interested early in philosophy of technology in North America tended to come from Marxist, phenomenological, and theological backgrounds, as well as from the older versions of pragmatism. Thus, in addition to an age difference, there was also a parentage difference.

The split in parentage, of course, paralleled the much wider gap which for many years was seen to form at the English Channel and which covered topics much wider than either the philosophy of science or the philosophy of technology. Unfortunately, accompanying this division were also habits of reading. The earlier dominant traditions in the philosophy of science remained ignorant of—or worse, hostile to—the strands of philosophy of science in Euro-American philosophy. By contrast, much Euro-American philosophy often appeared either uninterested in or hostile to science itself. This antipathy was, however, matched by strong interests in social concern for and critique of technological culture. Euro-American philosophy retained a strong interest in the ethical, social, and political implications of science and technology.

Underlying both these differences is the most profound difference of all. It is the difference of philosophical interpretation between what could be called the theory-prone and the praxis-prone approaches to both philosophy of science and philosophy of technology. But this difference is one which does not, at first, take the forefront.

One may thus speak of three gaps in the reading practices of those who should be interested in this interface between philosophy of science and philosophy of technology. The first gap is the one between philosophers who read primarily in one, and not the other specialization. The second gap crosses both fields in that some philosophers tend to read solely in either the Anglo- or Euro-American traditions without crossing style boundaries. And the third, even broader gap, lies between those who emphasize a theory-bias as opposed to a praxis-bias in the interpretation of the two specializations.

And while it certainly is true that in the last few years there have increasingly appeared both philosophers of science and philosophers of technology who might be called bi- or even multi-lingual with respect to the various tribal languages, there is not yet anything like a *lingua franca* (which, in science and technology probably would more likely be a *lingua inglesa,* anyway). Even as recently as the convergence of the "school" of instrumental realists whose books appeared from 1972 to 1987, there is little cross-citation between the first of the two reading-gap dimensions.

It is the third dimension of the reading gap which complicates the issue even further. This is the gap between what could roughly be called the theory-biased and the praxis-biased readings of both philosophy of science and philosophy of technology.

Philosophy of science, particularly pre-Kuhn, could have been characterized as dominantly located on one side of each of the three gap dimensions: It was interested almost solely in *science*—indeed, it remained largely blind to technology, even science's technology; it was largely Anglo-American (at least, after the U.S. immigration of the positivists), and it was theory-oriented. By contrast, philosophy of technology, arising from Euro-American concerns, was more interested in technology as a social-political phenomenon. It often left (pure) science in the background, and it dealt with a practical, even applied focus.

This was so much the case that, in the early days of what was to become the Society for Philosophy *and* Technology, the leadership deliberately resisted any moves to form the loose-knitted group into anything like its counterpart, the Philosophy of Science Association. That tendency remains to a certain extent, and there is no organization which actually calls itself a philosophy *of* technology organization. There are, however, publications which claim that title, including several books published as introductions to the philosophy of technology, such as Frederick Ferré's contribution to the Foundation Series—*Foundations of the Philosophy of Technology* (1988).

In 1979 I first entered this field with a book, *Technics and Praxis* (Reidel). It was the first volume in the Boston Philosophy of Science series dedicated explicitly to the philosophy of technology. The first half of that book focused upon what I called the *technological embodiment of science, its instrumentation.* Part of the thesis was that a crucial difference between modern and ancient science lay in its technology, its instrumentation. I still believe that is the case. I was, in effect, a philosopher of technology who, from the outset, was interested in the implications of technology *for* science. Moreover, coming from a tradition of praxis-interpretation I was also primarily concerned with epistemological and ontological issues. It was here that I saw the need to *interface* philosophy of science and philosophy of technology.

What I have set out to do responds to each of these facets of the philosophy of science/philosophy of technology interface. First, by underlining the roles of technology and its philosophy, I am seeking to elicit from neglected areas of the newer forms of philosophy of science a need to consider science's technology. But I am also seeking to do this by introducing from neglected Euro-American sources those important themes which are relevant to the philosophies of science and technology. As a tactic, I have gone back into fairly recent history and have done a brief re-reading of Thomas Kuhn as the progenitor of a "revolution" in the philosophy of science. Such recognition of Kuhn as a revolutionary is not new—but I have added several aspects to this re-reading. First, I *parallel* Kuhn with a European contemporary, Michel Foucault, who did for the social or human sciences something which turns out to be strikingly similar to what Kuhn did for natural science.

But in this identification of a proximity, I will begin to make thematic a certain perspective upon both Kuhn and phenomenology. The theme in question is that of a praxis-perception model of interpretation. The tactic I am about to take eventually *pairs* two strict contemporaries, Michel Foucault and Thomas Kuhn, both of whom published their first revolutionary works in the early sixties (Foucault in 1961, Kuhn in 1962). In situating this particular "take" on both Kuhn and Foucault, I am quite aware that the praxis-perception model I am trying to elicit varies from some interpretations of both phenomenology and Kuhn.

Second, I background both Kuhn and Foucault with the emergence of a view of science centering around praxis and perception, in the persons of Edmund Husserl and Maurice Merleau-Ponty. Lastly, I look at the rise of philosophy of technology alongside these figures in the thought of Martin Heidegger. It was the latter who raised the question of technology with respect to science and thus opened the way to an *interface* between the two subdisciplines.

I have focused the interstice between the two subdisciplines on the need to make technology—*instrumentation* in particular—focal as the point where science in its technological incarnation takes unique contemporary shape. I thus read the philosophy of science through its need for and neglect of a concern for technology.

Finally and most thematically, I outline what can be called an implicit *praxis-perception model* for both philosophy of science and technology. This model is explicitly derived from certain directions within phenomenology, but in a postfoundationalist guise. Here it serves as an interpretive guide for re-reading the role of instrumentation and technology in science.

To see more fully how praxis and perception play a crucial role in the way a technologically embodied science discovers knowledge, I have added to the recent histories of the philosophy of science and technology a brief look at five North American philosophers who have explicitly recognized this intersticial role: Hubert Dreyfus, as the background pioneer; Patrick Heelan, representing the Euro-American side; Robert Ackermann and Ian Hacking for the Anglo-American side; and myself. Although the traditions from which the two sides draw yield different results, it is clear that the roles of embodiment and instrumentation become crucial for both the philosophy of science and of technology.

As a tactic, I focused upon five authors whose *books* could be read in parallel, noting how they converge and form areas of consensus around the themes of what I call *instrumental realism.* However, this emergence of a consensus regarding the role of the technological embodiment of science's mode of knowledge gathering has, in recent times, taken a more complex turn. Today's instrumentation is rarely simple tools or measuring devices, and science as an institution is a corporately large enterprise. Both of these conditions complicate the result of a scientific process.

In the '80s, this complexity began to draw attention, particularly from sociologists and even anthropologists of science. By 1987, several books of importance focused specifically upon the way experiment relates to the "construction" of scientific realities and experiment as a socio-historical form of science. Here, too, is a genre of new approaches to the philosophy of science which merits independent treatment. But there are two points at which the praxis-perception approach taken here crosses over the social construction of scientific reality approach.

Our instrumental realists all began to realize that there were areas of ambiguity between instrumental "artifacts" (effects produced by instruments) and the related ambiguities in the complex technologically *produced* entities which constitute so much of science's interest today. Thus, a brief look at *experiment* was also called for, and following the simplified tactic of instrumental realism, I have looked at only the two most relevant books, by Peter Galison and Bruno Latour, concerning this issue. I conclude with broad observations on the effects of technology upon our wider existential world and on the interface which continues to be necessary between the philosophies of science and of technology.

I should like to give a few hints for the reader: At one level, this is a selective report on a few issues which arise in the state of the art in both subdisciplines. The earlier chapters are thus more historical and retrospective, in contrast to the strictly contemporary exposition in chapters 4, 5, and 6. Thus, chapters 2 and 3 belong together as a background in the interface, while chapters 4, 5, and 6 are more narrowly issue-related at the interface concern with science's instrumentation. I did debate about how to deal with my own works in that context, so I simply have shamelessly included myself as one of the five instrumental realist authors (I hope this slight break with some writing traditions will be forgiven). Obviously, one of the hopes of this sort of book is that the reader will use the occasion to go on to read the suite of books discussed. I did stick to one book each from my five selected authors, to make the matter simpler.

Similarly, in the chapter on experiment, I have focused upon only two authors whose books are most relevant to the themes discussed here. Such a tactic is obviously restrictive and yet suggestive. In the last decade both philosophy of technology and philosophy of science as fields have canonically exploded. *Instrumental Realism* is but one way of cutting across that exploded field and does not pretend to be either a survey or encyclopedic. It does claim to isolate themes which continue to be of serious issue within both philosophy of science and philosophy of technology.

Serious work on this book began during a sabbatical in the fall of 1984, although my interest in the topic and its focus upon a science/technology interface goes back another decade. There was a hiatus between that beginning and the completion of the manuscript in the fall of 1988 while I was on a research leave from my first term as dean of Humanities and Fine Arts at Stony Brook. That second time off mended the rent of the previous delay, so I am grateful to the university and its administration for supporting the writing.

I shall not draw up the usual long list of appreciated critics and readers, except for thanking both John Compton and Frederick Ferré for reading the first half of this book and making suggestions. Were it not for their agreement, I might not have proceeded to the conclusions

I illustrate here. I also thank Robert Ackermann and Joseph Rouse for their reading of the manuscript after its revision. Both gave insightful criticisms which I have tried to incorporate in the final draft. And I want especially to thank Steve Goldman and his colleagues at Lehigh University where, as a Selfridge Visiting Professor in the fall of 1989, I was able to enjoy their critique of the manuscript prior to publication. I must also add that Jean Kelley did her usual efficient work on manuscript preparation to which she always adds her initial editing.

Finally, while the "school" of instrumental realists I have pointed out arose largely independent of the first published books, there now has been some initial and fruitful interchange among some of its members and those who today work with experiment. I hope that one result of this book will be a furtherance of those initial exchanges.

Part One

Philosophy of Science Read through Philosophy of Technology

I. Introduction: Philosophers and Technology

Although there is a vast literature concerning technology, rarely has it been the primary theme of philosophers. But even with the plethora of books concerning the human impact of technology, few are concerned with the nature of technology per se. Either the literature tends to focus upon *effects* of technology or it may itself be technical literature. Rachel Laudan, one of the new group of historians concerned with the history of technology, has observed, "For all the diatribes about the disastrous effects of technology on modern life, for all the equally uncritical paeans to technology as the panacea for human ills, the vociferous pro- and anti-technology movements have failed to illuminate the nature of technology."[1]

Such a lack of focus also occurs within the disciplines, which again tend to focus either upon social effects or upon techniques of decision or management within the technological context. Again Laudan:

> On a more scholarly level, in the midst of claims by Marxists and non-Marxists alike about the technological underpinnings of the major social and economic changes of the last couple of centuries, and despite advice given to government and industry about managing science and technology by a small army of consultants and policy analysts, technology itself remains locked inside an impenetrable black box, a *deus ex machina* to be invoked when all other explanations of puzzling social and economic phenomena fail.[2]

What Laudan misses to a degree is that there has been some careful attention to technology by European philosophers, although that literature is not well known in the dominant North American circles.

The first book to coin the term "philosophy of technology" was written in 1877 by the neo-Hegelian Ernst Kapp. Of course, the later Marxian traditions also recognize the role of technology within human history and economics; but even more broadly, almost every major European philosopher in the dominant traditions of the mid-century also devoted attention to technology. Here the list would have to include Ortega y Gassett, Karl Jaspers, Nicolas Berbyaev, and Gabriel Marcel. But the name of Martin Heidegger would take the central position. Yet until recently, even Heidegger's clearly focused thematic concern with technology has gone largely without notice or comment from even the continentally oriented American philosophers within the scholarly traditions which follow this tradition here.

A second source of philosophical interest, one would think, should arise in the closest-related special interest among philosophers—the philosophy of science. This has not been the case until very recently, and then only after reaction, in effect, to challenges by minority groups within the profession of philosophy.

In a discussion of the state of the art, I shall turn to that often non-benign neglect for two purposes: First, I wish to locate the sources for a potential reevaluation of technology with respect to the philosophy of science; and second, I wish to use, by way of introduction, some of the more important European sources which show the greatest promise of development vis-à-vis an emergent philosophy of technology.

There has been for many years a thriving North American establishment in the philosophy of science. Indeed, the Philosophy of Science Association (founded in 1934) is the largest special-interest society within the profession of philosophy in America. The counterpart organization for technology, the Society for Philosophy and Technology, was not formally organized until 1983 (although it had met in a series of conferences for at least a decade before), and its membership is still less than a fifth of that of the PSA. This disproportion in organizational size is at least indicative of the degree and relative lateness with which philosophers in North America have come to a serious concern with technology.

There have been some beginnings: The major journal for the history and philosophy of technology, *Technology and Culture,* presented an issue as early as 1966 on "Toward a Philosophy of Technology." Among those who contributed were: James Feibelman, Mario Bunge, and Joseph Aggasi, who are still active in the field; and the organ of the Society for Philosophy and Technology, *Research in Philosophy and Technology,* which owed its primary existence to the work of Paul Durbin. Its reappearance is now under the hand of Frederick Ferré. However, *Research* collects articles which are usually quite diverse and interdisciplinary, and has not published either monographs or thematic books. But in contrast to the large European literature on the subject, there has been little serious work on the subject. As late as 1979, Mario Bunge could say:

> . . . Technophilosophy is still immature and uncertain of its very object, and does not exploit the entire scope of its own possibilities. That it is an underdeveloped branch of scholarship is suggested by the fact that so far no major philosopher has made it his central concern or written an important monograph on it.[3]

Bunge himself outlines, then, what he hopes will be the fields of "technophilosophy," his term for philosophy of technology. And although he was certainly aware that there were European strains, his attitude toward them reflects the negative attitude that most mainstream old philosophy of science has taken:

> Characteristically, such writers as Berdyev, Ellul, Heidegger, Marcuse, and Habermas fail to distinguish technology from its applications, and endow it with an autonomous existence and, moreover, power over man. [and, presumably referring to these same authors] . . . I do not count the tiresome tirades on the way technology "dehumanizes" man or robs him of his "authenticity"; this is not philosophy, but bad literature.[4]

But Bunge himself, along with the others mentioned (and to which one might add Edward Ballard, Peter Caws, Joe Margolis, and Marx Wartofsky as widely known philosophers who have written early articles on the subject), stands out from the more prevalent philosophical indifference to technology as a theme.

I cite this brief and recent state-of-the-art observation because it points to much deeper roots. There are persistent reasons to be found in the dominant traditions of the philosophy of science, which themselves form the context of vision which overlooks technology. In *Technics and Praxis*—written in 1979, Bunge's crucial year—I claimed:

> Part of the silence concerning technology comes from within philosophy itself. Philosophy usually conceives of itself more as a type of "conceptual" engineering than as a "material" engineering. Here there is a deeper set of relationships between science and technology as they emerge both in ancient and contemporary thought in philosophy.
>
> This symptomatology points to the dominance of a long "Platonistic" tradition with respect to science and technology, a tradition which, with respect to science and technology, turns out to be "idealistic." This conclusion turns upon the variable which I have called the primacy of praxis and is related only partly to the long-held distinction between theory and practice.
>
> The theory-practice distinction, however, may also be associated with a much deeper distinction, the mind-body distinction. Theory, as a set of concepts in some system of relations, is usually thought of as the product of mind, while practice often is associated with a product of body. And in the "Platonistic" tradition, mind takes precedence over body. Praxis philosophies return to this tradition in a new way because the primacy of a theory of action is one which positively evaluates what I shall call the phenomena of perception and embodiment.
>
> Contrarily, a "Platonistic" tradition is one which negatively judges, or at least evaluates, perception and embodiment as lower on the scale of human activity than what is presumed to be a "pure" conceptuality.[5]

The long tradition of Platonism noted here continues into the later concerns of the philosophy of science. It also determines the dominant view concerning the relationship between science and technology, which is that technology is *applied science,* or a merely neutral development from science.

Today, such a view is increasingly being called into question, perhaps most often by social scientists and historians, but also by philosophers, including many who are now working in the philosophy of technology. Again I cite my earlier *Technics and Praxis:*

> If one assumes that technology is an extension of science, a mere application and its instrument, then to address the effects of technology is at most to address a tertiary phenomenon. A series of relations may be formalized thusly:
>
> Science → technology → social effect
>
> Here the original cause is science as concept; technology is its effect or application; and the ethical or social effect is the tertiary phenomenon resultant from the series. Given this schematism, the only radical way of treating any problems which arise at the end of the series as other than symptoms would call for revision or change in the cause—in this case, the conceptual foundations of science itself.
>
> Of course, some philosophers do precisely this. The current debates about 'value free' science, even among neo-positivist philosophers, and the attacks upon various forms of scientific reductionism are, at least within the limitations of the idealist interpretation, working at the right level. The intuition that negative results from technology lead back to possible flaws in conceptual science is at least a consistent position with respect to Platonism.[6]

Since this now decade plus beginning, there has been an accelerated interest in the philosophy of technology. Here the principals have been thinkers like Langdon Winner with *The Whale and the Reactor* (1986), Albert Borgmann with *Technology and the Character of Contemporary Life* (1984), and even my own more recent *Technology and the Lifeworld* (1990). But in these cases the breadth of concerns is such that a focus upon the interface with the philosophy of science is often only sporadic within the books as a whole.

However, in comparison to the centuries-old traditions of Modern Science, philosophy of science is something of a relative newcomer. Prior to the nineteenth century, for example, it was often difficult to distinguish philosophy *from* science. What we would roughly call natural science today was, in early modern times (seventeenth to eighteenth centuries), *natural philosophy.* And even earlier it was, in fact, hard to distinguish science from its applications and embodiments, a fact which will have some bearing upon precisely the issue of the relation of science to technology. But with the gradual separation of science and philosophy until its much more extreme distinctions in professional form today, there was a more highly contrasted science-technology distinction, which eventually led to the disappearance of concerns with technology from philosophy.

To the informed, it would be unnecessary to note that the dominant philosophy of science in North America was first positivistic and then modified into what today is broadly called analytic philosophy. Both versions of philosophy of science were highly Platonistic in the sense sketched above; they conceived of science as a body of propositions, a conceptual, rational system, essentially disembodied from both social and material connections. This perspective views science as essentially a system of concepts and logical connections motivated by both explicit and implicit rational processes. This view has had a deep and profound effect even beyond the work of what I shall now call the *old* philosophy of science. For example, the sciences' own interpretations of themselves and their earlier history largely remained sympathetic to the positivist strain which continued to dominate early twentieth-century philosophy of science.

The father of the North American revolution in history and philosophy of science observed in the very opening chapter of *The Structure of Scientific Revolutions:*

> Even from history . . . [a] new concept will not be forthcoming if historical data continue to be sought and scrutinized mainly to answer questions posed by the unhistorical stereotype drawn from science texts. Those texts have, for example, often seemed to imply that the content of science is uniquely exemplified by the observations, laws, and theories described in their pages. Almost as regularly, the same books have been read as saying that scientific methods are simply the ones illustrated by the manipulative techniques used in gathering textbook data, together with the logical operations employed when relating those data to the textbook's theoretical generalizations.[7]

This is the attitude which accepts the notion of the older philosophy of science, an ideal and abstract science which apart from being ahistorical, is *disembodied.* It is a science without *perception* and a science without *technology.*

The same view, in the older historical disciplines, continued this implicit Platonism, which had as its implication the dominant view that technology must be the stepchild of science. Edwin Layton noted that historians, "while correctly repudiating the Marxist thesis that the Scientific Revolution was no more than the systematization of the knowledge of the craftsman, overreacted when they came to the converse conclusion, namely, that science was prior to and generative of technology."[8]

Given the usual conservatism of philosophical change, particularly within the mainstream of recent North American thought, it is hardly surprising that the few philosophers who have begun to develop interest in the philosophy of technology continue to carry over to it the same interests and assumptions which motivated their primary interest in science. Bunge's approach is symptomatic. In his article "The Five Buds of Technophilosophy," one can easily see this focus. He asks:

- Is there a technological method parallel to the scientific method and, if so, what are its rules and what is the efficiency of the latter?
- Some philosophers have claimed that, unlike science, technology has no laws and theories of its own. True or false? And, if true, what distinguishes technological law statements and theories from scientific ones?
- What are the peculiarities of the rules of advanced technology vis-à-vis mathematical and scientific rules?[9]

Here we have a virtually stipulative concept of technology, one which, by definition, illustrates the predisposition to interpret technology as applied science. In this case, vast areas of technological phenomena are effectively excluded—certainly traditional technologies and all variants other than those related to science. Moreover, as the historians have recognized, even many technologies which have been developed in the recent scientific milieu are not reducible to applications of science.

> When philosophy does turn its attention to the insistent presence of technology, it inevitably casts the question in one or another of the dominant modes of philosophical interpretation and reconstruction. Thus the *logic* of technological thinking and practice . . . and the question of technology's relation to science has been posed in the framework of the nomological model of explanation in the sciences—e.g., are there "laws" of technology or how does technology fit within the context of justification which defines the project of a logical-empiricist philosophy of science?[10]

But this is the *old* philosophy of science, regardless of its continued institutional power.[11] And even if, as we shall note, some of the preferences for purely conceptual concerns remain intact even in the *new* philosophy of science, a crack has appeared through which other possibilities may be glimpsed. What I call the new philosophy of science not only has opened the way to a different perspective upon science and its development but, in that very perspective, makes way for a philosophical concern with technology.

It is interesting that the roots of the new philosophy of science, particularly in the Anglo-American sector, are those which relate in every instance to historians and philosophers who have been sensitive to the historical, perceptual, and community embodiments of science. And whereas I shall turn shortly to some of the major philosophic aspects of this change of perspective, it should be noted that a good deal of separately originated and motivated work has also been carried out in the history of technology.

For example, with respect to the dominant notion of the science-technology relation, Rachel Laudan observes that the younger generation of historians of technology has become increasingly revisionist and has called this concept into question: "No less than three special issues of 'Technology and Culture' in recent years, not to mention countless individual articles and books, have been devoted to revisionist examinations of the relationship of science and technology."[12] This attack comes from both philosophical and more concretely, historical examination:

> Recent attacks on the concept of technology as applied science have employed two strategies, one empirical and one analytic. On the empirical front historian after historian has chronicled episodes in the development of technology where the major advances owed little or nothing to science. Whether one takes steam power, water power, machine tools, clock making or metallurgy, the conclusion is the same. The technology developed without the assistance of scientific theory, a position summed up by the slogan "*science owes more to the steam engine than the steam engine owes to science.* [Italics mine]"[13]

I am, of course, still characterizing the Anglo-American state of the art which has in its own way and by its own insights become revisionist. But in the wider world, it is not true that philosophers have failed to appreciate the insight of the historian's slogan. The smaller community of North American philosophers of science who are aware of the European traditions would immediately recognize in this questioning of the relationship an echo of Martin Heidegger's much earlier revolution in perspective. In the early fifties he not only had already made the question of technology a central one to his thought but had suggested a radical *inversion* of the science-technology relation:

> Because the essence of modern technology lies in enframing, modern technology must employ exact physical science. Through so doing the deceptive illusion arises that modern technology is applied physical science. This illusion can maintain itself only so long as neither the essential origin of modern science nor indeed the essence of modern technology is adequately found through questioning.[14]

In short, without returning to the presumed Marxian concept that the Industrial Revolution is a systematization of handcraft work, Heidegger asserts the reversal of Platonism. If anything, science becomes for

Heidegger a *technology*-science relation. I shall return to this theme later, noting only that the radicalness of this shift in perspectives has remained largely invisible to the old philosophy of science; and although not often explicitly noticed, it relates more closely to the perspective of some of the new philosophy of science. Had the Heideggerian counterpart to Platonism been a livelier part of the debate, perhaps it might not have taken so long for the new perspective to emerge.

What I call the old philosophy of science, however, remains largely unaware of and uninterested in even science's necessary technology, its instrumentation. Often satisfied that there is a sharp, or at least identifiable, distinction between "theoretical" and "experimental" science, the old philosophy of science has addressed itself almost exclusively to the former. This is partly in keeping with its own internal traditions, which view philosophy as an almost exclusively logical and linguistic exercise. Only recently have these prejudices begun to be called into question and revised in any serious way.

Yet the challenges which form what I shall call the "new" philosophy of science are irreversible, and they have begun to bear fruit for the interface between the philosophy of science and the philosophy of technology.

II. The New Philosophy of Science

The new philosophy of science arises, in part, out of explicit dissatisfaction with the disembodied and essentially idealistic and abstract notion of science often found within the dominant traditions of the old philosophy of science.

The new philosophy of science owes its emergence to a number of thinkers—or perhaps equally, to a change in perspective which arose independently in many quarters. I have already suggested that the shift which occurred to create the new philosophy of science had as one of its components a certain taste for the concrete embodiments of science. Again, limiting mention first to Anglo-American sources, one may note the impact of Karl Popper and Imre Lakatosz, who resituated science within its community of mutually interested persons, while focusing upon the emergence and development of science. The work of Stephen Toulmin should be noted as well. One might also mention the important work of Michael Polanyi, who introduced the notion of *tacit* knowledge to the traditions of the philosophy of science. Tacit knowledge was clearly an appreciation of one dimension of *praxis* and perception, a kind of "bodily knowledge" which plays a subterranean role even in science and which gradually began to be appreciated by some philosophers of science. But most scholars would surely agree that in the Anglo-American traditions, the person who most gave form to the new shape of philosophy of science was Thomas Kuhn.

I do not wish to go over the same ground already well trod since the publication of *The Structure of Scientific Revolution* (1962). The arguments and counterattacks, the extensions and revisions, have been many. But I shall take a different approach. I shall not address the mainline reaction, which accused Kuhn of both reducing science to a sociology of science and of exhibiting a presumed irrationalism, which arose from his shift of emphasis on discovery from its previously "purely" rational basis in forms of reasoning. Rather, what I see happening in Kuhn is the hint at a different model of interpreting science, a model which includes at least *perceptual* and, insofar as Kuhn is historically sensitive, *praxical* features often left out of standard accounts.

At the same time, I want to resituate Kuhn with respect to his proximity, probably unknowingly on his part, to phenomenology. My use of perception is closest to Merleau-Ponty's in the sense that sensory, or *bodily*, perception is always understood to be situated within a kind of cultural or contextual perception. Thus, while Kuhn, in a more Wittgensteinian sense, may often be using perception metaphorically as a kind of "intellectual" perception, here I am noting that what is sometimes taken as metaphorical is more than metaphorical.

Kuhn does not stand alone in what I shall describe as the *praxis-perception* model implicit in the new philosophy of science. Until recently, beyond the gaze of many philosophers, there was a parallel development of an amazingly similar impact made upon the human sciences by the work of the late French intellectual historian-philosopher Michel Foucault. His tutor, usually unmentioned but all-too-clearly visible in his very invisibility, was Maurice Merleau-Ponty, the philosopher of perception par excellence. In turn, Merleau-Ponty draws from an even older development of a praxis-perception model of interpretation outlined in the work of the later Husserl. I shall examine this parallelism.

This examination will take the shape of a chronologically discontinuous exposition which, while beginning with Kuhn, then reverts to an earlier period beginning with Husserl and leading back up to Kuhn's contemporary, Michel Foucault. I do this first with respect to philosophy of science but with deliberate focus upon the often indirect roles of technology in these contexts.

First the Kuhnian revolution: The critical, or negative, thrust of Kuhn's reinterpretation of scientific development was immediately felt. It could easily be construed as a direct attack upon the analytic-positivistic-nomological model of science. For Kuhn, the laws of science, the rules of operation, the system of induction and deduction—while in no way ignored or rejected—were undercut as foundational. More basic to the operation of science was a *paradigm*. In effect, what Kuhn's negative attack accomplished was an inversion of priorities, making the nomological model of science derivative.

Kuhn first described his notion of paradigm in a highly general form, stating that "a paradigm is an accepted model or pattern"[1] which guides the development of *normal* science. At first, it might be thought that such a model was merely the particular arrangement of parts of a theory, itself a kind of higher order, but perhaps an implicit "rule."

> In this standard example, the paradigm functions by permitting the replication of examples any one of which could in principle serve to replace it. In a science . . . a paradigm is rarely an object for replication. Instead, like the accepted judicial decision in common law, it is an object for further articulation and specification under new or more stringent conditions.[2]

But a paradigm is *presupposed* by the operations of normal science, which is to say that what constitutes the paradigm is also more basic as a condition of the possibility for normal science. Laws, rules, the nomological model become not founding, but founded. "Perhaps it is not apparent that a paradigm is prerequisite to the discovery of laws . . . ,"[3] but the relation is soon made explicit. "Rules, I suggest, derive from paradigms, but paradigms can guide research even in the absence of rules."[4] Rules become twice secondary. Paradigms are the very means by which theory can operate: "Paradigms provide all phenomena except anomalies with a theory-determined place in the scientist's field of vision."[5] They are ultimately the ground of normal science itself: "Without commitment to a paradigm there could be no normal science."[6]

And while recognizing that normal science is also the dominant way in which science operates, it too is inverted with respect to its grounds. *Revolutionary* science, which occurs through a paradigm shift, is more basic to scientific development or advancement. Normal science, such as that portrayed in the textbooks—or equivalently, by the nomological model—is sedimented science. It is carrying out the refinements and extensions of some previously adapted paradigm.

One can immediately see why, in its critical dimension, *The Structure of Scientific Revolutions* became controversial. In a subtle sense, however, the book became its own revolutionary fulfillment, itself a paradigm shift in the interpretation of science—so much so that today its perspective is itself virtually "normal." In fact, the response within large segments of the scientific community was such that the very language of Kuhn began to be used by scientists in their self-interpretations.

However, the established philosophy of science community was less enthusiastic. Kuhn was frequently dismissed as a mere "sociologist of knowledge," or worse, an *irrationalist,* since his model of scientific change clearly included factors other than sheer rational calculations or logical connections. Thus, even though adapted by a now sizeable younger generation of both historians and philosophers of science, the persistence of the old philosophy of science remains. Philosophies, in my estimation, rarely die, even if refuted or undercut. More likely, they either go underground—reluctantly yielding even the smallest terrain—or, more frequently, resuscitate themselves in a new guise. I even suspect that the enduring, institutional aspects of philosophy are not that different from what happens in other forms of human industry. The historian Edward Constant once observed that in the transition from piston-engined to turbojet aircraft:

> Old communities and traditions virtually never give birth to radically new technologies. No manufacturer of piston aircraft engines invented or

> independently developed a turbo-jet. No designer of conventional reciprocating steam engines invented a steam turbine, no manufacturer of steam locomotives independently developed diesel engines. In the case of both firms and individuals, community practice defines a cognitive universe that inhibits recognition of radical alternatives to conventional practice.[7]

This observation on industry appears to me to apply equally well to philosophical establishments where the development of new fields or approaches is concerned!

The critique, however, is only the negative side of the new philosophy of science. Its positive side is the emergence of what I shall call a *perceptual model* of interpretation. Kuhn himself makes this point repeatedly:

> Examining the record of past research from the vantage of contemporary historiography, the historian of science may be tempted to exclaim that when paradigms change, the world itself changes with them. Led by a new paradigm, scientists adopt new instruments and look in new places. Even more important, during revolutions scientists see new and different things when looking with familiar instruments in places they have looked before.[8]

Kuhn characterizes this specifically as a *way of seeing.* It is what I shall call, in this case, an example of *structured macroperception.* Kuhn astutely recognized that phenomena may be seen with different selectivities—selectivities which call into question whether the thing seen is similar in any way to that which was previously seen. His specific metaphor for such changes was the gestalt shift, such as occurs with ambiguous pictures (as with Wittgenstein's duck/rabbit; indeed, Wittgenstein is an important figure in Kuhn's background).

Kuhn repeatedly gives examples of such shifts, emphasizing *radical discontinuities* implied in such gestalt shifts. For example, a minor shift concerning stars and planets occurred between 1690 and 1781. Uranus was first identified as a star; then, after a switch of interpretation, as a planet. There followed the identification of numerous astronomical phenomena as planets rather than stars, to the extent that twenty were identified![9] But much more telling as a shift of vision is the following case:

> During the seventeenth century, when their research was guided by one or another effluvium theory, electricians repeatedly saw chaff particles rebound from, or fall off, the electrified bodies that had attracted them. At least that is what seventeenth-century observers said they saw, and we have no more reason to doubt their reports of perception than our own. Placed before the same apparatus, a modern observer would see electrostatic repulsion (rather than mechanical or gravitational rebounding), but historically . . . electrostatic repulsion was not seen as

> such until Hauksbee's large-scale apparatus had greatly magnified its effects.[10]

Note, anticipatorily, that there is a relation here between perception and instrumentation (technology). Kuhn suggests that the use of the same instrumentation can give rise to differing perceptions, but *historically* the shift did not occur until the instrumentation itself changed. This is not without import, but it remained only an underdeveloped background phenomenon within Kuhn's approach.

For Kuhn, gestalt shifts are shifts in *seeing as* (Wittgenstein). What is explicit in his interpretation are such things as changes in what counts, selectivities within the phenomenon: "for example . . . when Aristotle and Galileo looked at swinging stones, the first saw constrained fall, the second a pendulum."[11] Indeed, only through a shift could pendulums be perceived: "Pendulums were brought into existence by something very like a paradigm-induced gestalt shift."[12] A shift of perception radically reorganizes not some particular element, but a whole field. "Paradigms determine large areas of experience at the same time."[13] This analysis of scientific perception, once understood, makes it easy to see why the nomological model must take a different role within the interpretation of science. Only after there is some paradigm or other, some structure macroperception or other, does what counts as a fact become a fact. Similarly, only after there is a gestalt can its laws be determined and refined. The same can even be said for predictions—only after there is some formed whole (gestalt) can there be anything like a rational prediction. As such, a perception is a kind of precursor to a phenomenon, which then can be made ever more explicit in its detail and implication. Kuhn's strategy is thus a "top down" one.

Within the context of Anglo-American history and philosophy of science, the Kuhnian adaption of a perceptual model was indeed daring. This was so not only because it cut against the grain of the then dominant strands of the philosophy of science, but because it did so in the absence of a much richer tradition of the interpretation of perception itself, a tradition which was much better known in European circles through early phenomenology (Husserl and Merleau-Ponty). Clearly, Kuhn did draw upon the later Wittgenstein, who, among all the philosophers associated with early analytic philosophy, perhaps most often used and referred to perceptual examples. Learning a language is closely linked to learning to perceive, as Kuhn clearly states:

> The child who transfers the word "mama" from all humans to all females and then to his mother is not just learning what "mama" means or who his mother is. Simultaneously he is learning some of the differences

> between males and females as well as something about the ways in which all but one female will behave toward him. His reactions, expectation, and beliefs—indeed, much of his perceived world—change accordingly.[14]

The Kuhnian model is, then, basically a Wittgensteinian perceptual model. It recognized that gestalts are perceptual contexts within which elements may be radically and discontinuously related, not only in terms of rearrangements, but more radically, in terms of items which drop out and disappear and in which others appear for the first time. This places Kuhn, with or without his knowledge or approval, very close to one strain within perceptualist phenomenology. Moreover, in his description of scientific perception he even approximates the notion of intentionality, which I have suggested forms the framework for the phenomenological interpretation of perception. If the "world" changes in a paradigm shift, the object or reference of perception within its entire field changes; it *reflexively* implies a change of some kind in the perceiver. In the structured macroperception of the scientific community, this also occurs: "Successive paradigms tell us different things about the population of the universe and about that population's behavior. . . . But paradigms differ in more than substance, for they are directed not only to nature but also back upon the science that produced them."[15] Or again, " . . . when paradigms change, the world itself changes with them . . . paradigm changes do cause scientists to see the world of their research-engagement differently."[16]

So, from within the traditions of Anglo-American history and philosophy of science there has arisen a perceptual model for the interpretation of scientific change and development. But is this model relevant to the counterpart philosophy of technology? There is at least the hint, within the context of the history of science, that it may be. Kuhn rarely emphasizes the role of technologies—which, for science, are usually its instrumentation—in any thematic way. But instrumentation, at least for all modern science, is the material condition; better, it is the *embodiment* by which science *perceives*. Here one finds at least one clue to a possible shift from the philosophy of science to the philosophy of technology. Yet this clue is not explicitly developed by Kuhn, who remains primarily theory oriented.

Before following up on this clue, however, I wish to turn to Kuhn's unrecognized philosophical kin, the early phenomenologists. For it was in this early twentieth-century movement that what I shall term the *praxis-perception* model was more fully developed. Thus, I am adopting this European tradition into the family of the new philosophy of science.

The primary progenitor of the phenomenological movement was Edmund Husserl, whose published works appeared from the early 1900s through 1936. A French reader of his, particularly with regard to the later works, was Maurice Merleau-Ponty, who died in 1961. Both turned their attention to the role of perception in science, and both utilized a version of the lifeworld as an interpretative idea.

Husserl's *Crisis* was published in 1936, and in it one can find a considerable parallelism with what was to become the new philosophy of science. Admittedly, in now rereading the *Crisis, aprés-*Kuhn, one cannot help but discern contrasts as well. First, Husserl continued the German tendency to view science as Wissenschaften—although his examples indeed surrounded the rise of Modern physics in Galileo and Descartes—rather than take the narrower focus upon natural science as exemplary science. Then too, his aim was in part to discern a unique rationality in Western thought; this colors his interpretation and gives it the appearance of a normative, if not "linear," development. And while, as I shall illustrate, he clearly develops an idea of something like paradigm shifts, he also regards each change as necessarily both a gain and a loss; thus, the question of progress becomes enigmatic (this in spite of Husserl's own hope for philosophical progress).

What I wish to concentrate upon, however, is the beginning of a praxis-perception model of interpretation. In Husserl, this notion takes its place within the structure of the *lifeworld.* Scholars have long acknowledged that there is a fundamental ambiguity within the idea of the lifeworld which will set the stage for the exposition of phenomenology to follow. On the explicit level, Husserl remains a foundationalist. This is to say that some stratum—in this case, of human activity—is a founding stratum, while others are founded, are dependent, upon the founding fundament. In Husserl's case, what is fundamental is a kind of ordinary human praxis and perception, the world of the human interaction among material things and others. Its openness toward the other is sensory, and this relation is focally perceptual.

This foundation in ordinary perception and praxis is, however, not usually critically examined; it is simply taken for granted. This ordinary context is what is both basic and shared throughout the human community.

> Consciously we always live in the lifeworld; normally there is no reason for making it explicitly thematic for ourselves universally *as* world. Conscious of the world as a horizon, we live for our particular ends, whether as momentary and change ones or as an enduring goal that we have elected for ourselves as a life vocation, to be the dominant one in our active life. . . . Thus as men with a vocation we may permit ourselves to be indifferent to everything else, and we may have an eye only for this horizon as our world and for its own actualities and possibilities—those that exist in this "world"—i.e. we have an eye only to what is in "reality" here. . . . [17]

Within this basic and universal domain of perception and praxis there may occur special forms of activity with particular selectivities which may become "sciences." Science in this concept will both be related to, but, in specific ways, distinguishable from, the lifeworld foundation in ordinary experience. The lifeworld may contain various "worlds." "The goal-directed life which is that of the scientist's life-vocation clearly falls under the generality of the characterization just made, together with the 'world' that is awakened therein in the communalization of scientists . . . as the horizon of scientific works."[18]

Here Husserl retains his usual lofty perspective but also anticipates the awareness that embodies science in its community and activity. It is clear that experience within the horizon of the world and the different "worlds" which can be constituted within the lifeworld is the framework by which development in science can occur. Returning again to "The Origin of Geometry," we find one slightly more specific hint of how such an analysis might take shape:

> Geometry and the sciences most closely related to it have to do with space-time and the shapes, figures, also shapes of motion, alterations of deformation, etc., that are possible within space-time, particularly the measurable magnitudes. It is now clear that even if we know almost nothing about the historical surrounding world of the first geometers, this much is certain as an invariant, essential structure that it was a world of "things" (including the human beings themselves as subjects of this world); that all things necessarily had to have a bodily character . . . and [these] can be secured at least in [their] essential nucleus through a careful a priori explication, [in that] these pure bodies had spatio-temporal shapes and "material" qualities. . . . Further, it is clear that in the life of practical needs certain particularizations of shape stood out and that a technical praxis always [aimed at] the production of particular shapes and the improvement of them according to certain directions of gradualness.[19]

To this point one may contrast several aspects of the lifeworld with the "worlds" of sciences. First, ordinary perception and action is primary and universal and is simply presupposed by the actual scientist. Second, the lifeworld may be said to include the "world" of science, but not vice versa. And third, there is a marked contrast between the "perceptions of the sciences and perception in the lifeworld." Here we continue to take account of the first sense of lifeworld as founding fundament.

The ordinary perception, examined critically and reflectively, is shown to be different from scientific perception. In his case studies on the origins of geometrical method, Husserl notes that:

> In the intuitively given surrounding world, by abstractively directing our view to the mere spatiotemporal shapes, we experience "bodies"—not

> geometrically-ideal bodies but precisely those bodies that we actually experience, with the content which is the actual content of experience.[20]

This perception is what I call *microperception;* it is more narrowly sensory in its original understanding. The base stratum of the Husserlian lifeworld continues to be this domain of human interaction with a material-surrounding world.

Geometry, when it does arise, does so from a particularization within the perceptual field: Certain shapes are noticed, preferred, perfected, etc., and by a gradual process of abstraction and variation, move from that base toward the imaginatively ideal. Geometry originates from a certain kind of perception and praxis:

> The geometrical methodology of operatively determining some and finally all ideal shapes, beginning with basic shapes as elementary means of determination, points back to the methodology of determination and measuring in general, practices first primitively and then as an art in the prescientific, intuitively given surrounding world.[21]

As these shapes are selected, chosen, and perfected, new interests and praxes arise with a trajectory toward idealization. "Out of the praxis of perfecting, of freely pressing toward the horizons of *conceivable* perfecting 'again and again' *limit-shapes* emerge toward which the particular series of perfectings end."[22] In short, one begins to get a special geometrical praxis.

In this new type of praxis, perceptions also change. A new praxis is an *acquisition,* which once acquired may become familiar; its origins and the means by which it was attained are forgotten. That which becomes familiar becomes transparent and taken-for-true. It becomes a kind of "perception," but now, while intuitive, something beginning to approximate a macroperception, a "cultural" perception. It is precisely this movement which characterizes Husserl's interpretation of the rise of Modern Science in the figures of Galileo and Descartes.

I shall not sketch out the whole of this interpretation. But what Husserl asks is: What constitutes the realm of the taken-for-granted which would have been part of Galileo's perspective upon geometry? And with some subtlety he traces not only what Galileo could take for granted but, in a reconstruction, makes us aware of what we take for granted. This level actually arises much later, after the paradigm shift Galileo instigated was solidified.

According to Husserl, what was "obvious" to Galileo was a long tradition of the relation and application of ancient geometry in a Platonistic guise such that the empirical world could be mathematized with a certain intuitiveness—but only to a point. Partial measurements and correspondences were known from antiquity (and revived in the

Renaissance). Thus, without further specifying origins, proportions between lengths of strings and sounds (harmonics) and between selected shapes (triangles, etc.—not initially rough or complex shapes) and their elaboration in plane geometry could be taken for granted. But a new type of universalization, a new perspective, is taken by Galileo, a perspective which only much later can become so taken-for-granted that it becomes intuitive.

The problem revolves around sensory perception. Shapes, already closely subsumed under ancient geometry, are only a part of the sensory world. In addition, there are what Husserl calls plenary or specific-sense qualities (colors will do for an example). These do not easily fall under the geometric praxis: "The difficulty here lies in the fact that material plena—the specific sense-qualities—which concretely fill out the spatiotemporal shape-aspects of the world of bodies *cannot,* in their own gradations, be *directly* treated as are the shapes themselves."[23]

Some means must be devised, then, to account for these plenary qualities, *or else* one must realize that the geometrical method is only a special and limited praxis related to one aspect of the world. Galileo's invention, Husserl claims, is the development of an *indirect* means of mathematizing the plena. Galileo must find a means to *translate* plenary phenomena *into* spatial ones in order for them to become available for geometrical analysis. And that is what he does.

Conceptually, this move, well known in both Galileo and Descartes, first denies to the objects their plenary qualities in the doctrine of primary and secondary qualities. Put baldly, the object-in-itself is purely a geometrical entity, a *res extensa;* its plenary qualities are located not in the object but in the subject. Colors are "subjective." But now, since we see a thing as both extended and colored, there must be some way to subsume color into geometrical analysis. And it is here that the *indirect* geometrization begins to take shape. There must be some index of regularity which, while not directly spatial, can be related to some "spatial" measurement, *directly* or through a process of *translation.*

> Now with regard to the "indirect" mathematization of that aspect of the world which in itself has no mathematizable world-form, such mathematization is thinkable only in the sense that the specifically sensible qualities ("plena") that can be experienced in the intuited bodies are closely related in a quite peculiar and *regulated* way with the shapes that belong essentially to them.[24]

In this perspective, not yet intuitive, Galileo dramatically paves the way for Modern physics such that, today, second thoughts are rarely given to the procedure:

> What we experienced in prescientific life, as colors, tones, warmth, and weight belonging to the things themselves and experienced causally as a body's radiation of warmth which makes adjacent bodies warm, and the like, indicates in terms of physics, of course, tone-vibrations, warmth-vibrations, i.e., *pure events in the world of shapes.* [Italics mine][25]

So much of this is taken for granted that undergraduates can even say that they "see" wave lengths.

This is to say that once gestalted, the Galilean perspective becomes a kind of macroperception which can be taken for granted with new modifications possible. But it is precisely here that the ambiguity in Husserl reaches its own apex. For *if* the new means of understanding phenomena becomes a genuine cultural acquisition as the investigation into the origins by means of a praxis become transparent, it overlooks the contrast with the fundamental lifeworld.

> But now we must note something of the highest importance that occurred even as early as Galileo the surreptitious substitution of the mathematically substructed world of idealities for the only real world, the one that is actually given through perception, that is ever experienced and experienceable—our everyday lifeworld. This substitution was promptly passed on to his successors, the physicists of all the succeeding centuries.[26]

Husserl seems to be saying that the lifeworld is, and must be, the sensory lifeworld, based in the relations between actional humans and the concrete, material world of things and beings, which are bodily. And these are intuitively, perceptually available to everyone. Then a second type of intuitability also occurs, such as that exemplified in the Galilean revolution, in which certain combinations of praxically attained perspectives can make possible another intuitable attainment, a cultural acquisition, like a science.

However, such an acquisition is also ambiguous. Because what is gained by the very means of mathematization, Husserl argues, loses an essential sense of concreteness by overlooking the fundamental lifeworld.

> Galileo, the discoverer . . . of physics, or physical nature is at once a discovering and a concealing genius. He discovers mathematical nature, the methodical idea, he blazes the trail for the infinite number of physical discoveries and discoverers. [But] immediately with Galileo, then, begins the surreptitious substitution of idealized nature for prescientifically intuited nature.[27]

The acquisition of the new paradigm conceals the fundament of the ordinary dimension of the lifeworld.

In this context, however, two remarks are called for: First, it should now be obvious that Husserl clearly developed something similar to the notion of a paradigm shift. This is, in essence, his interpretation of Galileo. Galileo "sees" the world differently, and once this perspective is established, it can itself become a tradition which persists. Here is normal science, a way of seeing and interpreting phenomena which contrasts with the world of ordinary activity and seeing. In this sense, and specifically by means of an experiential analysis, Husserl anticipates the new philosophy of science.

For Husserl, praxis and perception are the focal basis of the lifeworld. The lifeworld is that structure of experience which is both perceptual and historical. It contains *sediments* and *traditions* and what Husserl proposed in his reexamination of the rise of Modern Science.

Interestingly, at this very opening point, a long-persistent misunderstanding of Husserl and phenomenology has barred the way to seeing Husserl as forerunner to the new philosophy of science. It is thought that Husserl always began with "immediate experience," which was *intuited.* Therefore, the misunderstanding usually goes, phenomenology becomes a merely "subjective" procedure. What fails to be noted is that for phenomenology, *all intuitions,* as well as all sediments and traditions, *are constituted, not given.* Givenness is always merely indexical or preliminary.

Intuitions and traditions, when examined critically, are shown to be what they are only when their constituting field or context is made apparent; and that is what Husserl did with Galileo. "Given" intuitions or traditions are truly parallel to what Kuhn takes as normal science, while the establishing of a perspective, as with Galileo, is a revolution in science. Truth—the obvious, the transparent—is what it is because it is already familiar, constituted, sedimented. But this, in turn, must be referred to both the field which makes the constituting possible and to the activity which does the constituting (or, the revolution which establishes it).

The second comment relates to the tension between Husserl's two senses of perceivability which may be found in the idea of the lifeworld. Husserl himself retained the hard sense of perceivability for that founding dimension of the lifeworld or perception in its micro- (sensory) signification. The second or macro- (hermeneutic-cultural) signification was for him an aperceptive sense. But insofar as the second sense of perception is used and insofar as it could be gestalted into a taken-for-granted cultural tradition, it is clearly also "intuited." Furthermore, Husserl allows that there is both an ordinary and a mathematical praxis possible within the lifeworld. However, he was still a foundationalist in that the first, hard sense remained for him the founding stratum.

While Husserl retained microperception as foundational, making (specific scientific) macroperception derivative, even if "higher," Kuhn as the later counterpart to this observation emphasizes the second dimension of perceivability. For Kuhn, it is primarily macroperception which is foundational. In a rough sense, specific perceptions (scientific observations) take place within a paradigm or macroperception. Scientific perception, while clearly bodily-perceptual in some sense, is primarily that of the macro-order, a specific formation of the hermeneutic-cultural order.

Interestingly, the movement toward the more obvious discontinuities and polymorphous qualities of cultural macroperception also characterizes some of the moves of the post-Husserlians. It occurs dramatically in the work of Maurice Merleau-Ponty.

Within the phenomenological traditions, Merleau-Ponty was the preeminent philosopher of perception. His *Phenomenology of Perception* (first published in English in 1962) remains the classic study in the field and, in the context here, was the work which most thoroughly distinguished a phenomenological theory of bodily-sensory perception. His later *Visible and the Invisible* (first published in English in 1968), while a posthumous unfinished work, displays a modification which focuses upon the dimension of cultural perception or macroperception as I interpret it. My reference to him here, however, must be limited to taking note of the emergence of certain features of a praxis-perception model of interpretation relevant to the context of philosophy of science and technology.

Merleau-Ponty read and adapted Husserl into the French world, but his focus began with the later Husserl of the lifeworld period. In the earlier *Phenomenology of Perception* there is a kind of structural adaptation of the lifeworld notion. The base of the lifeworld, as in Husserl, could be said to be common experience:

> All my knowledge of the world, even my scientific knowledge, is gained from my own particular point of view, or from some experience of the world without which the symbols of science would be meaningless. The whole universe of science is built upon the world as directly experienced, and if we want to subject science itself to rigorous scrutiny and arrive at a precise assessment of its meaning and scope, we must begin by reawakening the basic experience of the world of which science is the second-order expression.[28]

There is thus a distinction, similar to Husserl's, between a common, primary relation to the world and a special, secondary (perhaps higher), scientific relation to the world. Phenomenology, as Merleau-Ponty sometimes characterized it, was the philosophy which must take the measure of this difference.

If the foundation of the lifeworld could be found in common experience, pre-critically it remains only that which is taken-for-granted. The task of a critical phenomenology was to awaken its sense—and here Merleau-Ponty also adapted a version of the Husserlian epochē:

> It is because we are through and through compounded of relationships with the world that for us the only way to become aware of the fact is to suspend the resultant activity, to refuse it our complicity . . . or yet again, to put it "out of play." Not because we reject the certainties of common sense and a natural attitude towards things—they are, on the contrary, the constant theme of philosophy—but because, being the presupposed basis of any thought, they are taken for granted, and go unnoticed, and because in order to arouse them and bring them into view, we have to suspend for a moment our recognition of them.[29]

So far, fairly standard Husserl—but Merleau-Ponty interprets the primary world-self relation (intentionality) as *actional* and *perceptual.* In defense of his thesis he claimed, "The perceived world is always the presupposed foundation of all rationality, all value, all existence."[30]Thus, in a complex and broadened phenomenological sense, he makes perception basic:

> I cannot put perception into the same category as the synthesis represented by judgements, acts or predications. We must not, therefore, wonder whether we really perceive a world, we must instead say: the world is what we perceive. . . . To seek the essence of perception is to declare that perception is, not presumed true, but defined as access to truth.[31]

Perception, in this sense, is access to truth, the limits within which even rationality can occur: "Rationality is precisely measured by the experience in which it is disclosed. To say that there exists rationality is to say that perspectives blend, perceptions confirm each other, a meaning emerges."[32]

By making perception foundational, Merleau-Ponty could be said to be something of a phenomenological "empiricist," even though his interpretation of perception radically differs from any empiricist sensation theory. However, by initially identifying perception with the foundationalist interpretation given it by Husserl, Merleau-Ponty falls into the same trap regarding science, which we have previously noted in Husserl's interpretation of Galileo:

> Scientific points of view, according to which my existence is a moment of the world's, are always both naive and at the same time dishonest, because they take for granted, without explicitly mentioning it, the other point of view, namely that of consciousness, through which from the outset a world forms itself around me and begins to exist for me. To return to things themselves is to return to that world which precedes knowledge, of which knowledge always speaks, and in relation to which every scientific

> schematization is an abstract and derivative sign-language, as is geography in relation to the country side in which we have learnt beforehand what a forest, a prairie or a river is.[33]

I shall contend that by relating common perception to scientific perception in this foundationalist way, the contrast between the praxis of ordinary existence and scientific activity is not only too sharp, but it overlooks another strategy entirely, a strategy which could have been developed from within the phenomenological perspective itself.

The point here, however, is to take brief account of the praxis-perception model which occurs in the works of Merleau-Ponty as a potential contribution to both the philosophy of science and the philosophy of technology. In general, the modifications upon Husserlian phenomenology have long been recognized as *existential* reinterpretations." [Phenomenology] far from being, as has been thought, a procedure of idealistic philosophy . . . belongs to existential philosophy: Heidegger's "being-in-the-world" appears only against the background of the phenomenological reduction."[34] Note that existential here means precisely the kind of "materiality" which occurs in praxis and perception; it is primarily an account of a world-body-as-me relation rather than the analysis of "pure consciousness" as intimated by Husserl.

This is to say that the I-world correlation, which is *intentionality,* is reinterpreted by Merleau-Ponty to be a correlation between the experienced environment—the world—and my experiencing of it as an incarnate or embodied being. This version of perception thus focuses upon bodily concreteness and revolves around the various dimensions of bodily existence. The immediate consequences of a theory of perception are clear. Not only is body (as experiencing or "lived" body) implicated in all perception and the condition of what and how something is perceived, but "the theory of the body is already a theory of perception."[35] Thus, in the explication of this theory, existential spatiality or bodily position must be constantly accounted for:

> When I walk around my flat, the various aspects in which it presents itself to me could not possibly appear as views of one and the same thing if I did not know that each of them represents the flat seen from one spot or another, and if I were unaware of my own movements, and of my body as retaining its identity through the stages of these movements.[36]

Here we return to one of the points made concerning phenomenological relativity. Whether or not it is interpreted foundationally, there is some kind of peculiar privilege of bodily existence with respect to any phenomenological perspective. At least, this is so for all microperception. Not to take account of spatiality (position), temporality (existential time), or the various dimensions of actional perception (mortality, expressiveness, even sexuality) would be inadequate and would dehumanize the account.

I shall briefly outline a few of the succinct features of the Merleau-Pontean theory of perception:

1) The recovery of lifeworld perception is understood by Merleau-Ponty to be the task of rediscovering the complexity and multidimensionality of perception. In the critical part of his analysis—and it should be noted that the science which he criticizes is primarily *psychology*—his objections include the various types of reductionism which characterize much twentieth-century human science. What is primitive in perception is its richness:

> A wooden wheel placed on the ground is not, *for sight,* the same thing as a wheel bearing a load. A body at rest because no force is being exerted upon it is again for sight not the same thing as a body in which opposing forces are in equilibrium. . . . [V]ision is already inhabited by a significance which gives it a function in the spectacle of the world and in our existence. . . . The problem is to understand these strange relationships which are woven between the parts of the landscape, or between it and me as incarnate subject, and through which an object perceived can concentrate in itself a whole scene or become the *image* of the whole segment of life. Sense experience is that vital communication with the world which makes it present as a familiar setting of our life. It is to it that the perceived object and the perceiving subject owe their thickness.[37]

The perceptual primitive is complex, multidimensioned. In this sense, the analysis is non-reductive.

2) Although the various senses are discussed throughout the *Phenomenology of Perception,* it is also clear that bodily motility, action, is basic. Not unlike much psychology, various illusions and perceptual deception fascinate Merleau-Ponty. He finds the clue to explanation residing primarily in full, bodily engagements with these (an illusion can maintain itself only through a certain distance and abstraction). In a discussion of the inverting-glasses experiments (glasses which presumably invert our "visual image" of the environment) and of distorted rooms (which look normal only at a distance and from one position but, if entered, are obliquely constructed and thus non-familiar), the illusory quality depends upon praxis and bodily action. "What counts for the orientation of the spectacle is not my body as it in fact is, as a thing in objective space, but as a system of possible actions, a virtual body with its phenomenal 'place' defined by its task and situation. My body is wherever there is something to be done."[38]

Bodily existence is both actional and *oriented.* Directions are not arbitrary but refer to bodily capacities and relations within potential tasks (thus up/down, right/left, and forward/back belong in oriented ways to a being of upright posture).

> We must not wonder why being is orientated, why existence is spatial, why . . . [the body's] co-existence with the world magnetizes experience and induces a direction in it. The question could only be asked if the facts were fortuitous happenings to a subject and an object indifferent to space, whereas perceptual experience shows that they are presupposed in our primordial encounter with being, and that being is synonymous with being situated.[39]

3) If body-world relations interpreted actionally are basic, the perceptual model which ties much of the Merleau-Pontean strategy together is that of the gestalt figure/ground phenomenon:

> When Gestalt theory informs us that a figure on a background is the simplest sense-given available to us, we reply that this is not a contingent characteristic of factual perception . . . it is the very definition of the phenomenon of perception, that without which a phenomenon cannot be said to be perception at all. The perceptual "something" is always in the middle of something else, it always forms part of a field.[40]

This is why, for a phenomenological analysis, there can never be an isolated thing-in-itself. All things are related to some context or other. This "invariant" for phenomenology, however, derives from its perceptualism.

In Merleau-Ponty's case, the gestalt model even relates to bodily motility in basic ways:

> My body is geared onto the world when my perception presents me with a spectacle as varied and as clearly articulated as possible, and when my motor intentions, as they unfold, receive the responses they expect from the world. This maximum sharpness of perception and action points clearly to a perceptual *ground*, a basis of my life, a general setting in which my body can co-exist with the world.[41]

4) If the basic model for analysis revolves around figure/ground relations, there is a related whole/part strategy which also governs the analysis. In the discussion of illusions, it again becomes clear that there is a weighted difference between the whole (world) and its parts (particular objects or perceptual takings). "There is absolute certainty of the world in general, but not of any one thing in particular."[42] These top-down and whole-part strategies are common to most phenomenological procedures.

5) Since within any given context or whole there can be uncertainty about parts, the human perceptual situation is both fluid and ambiguous. If we join Merleau-Ponty in accepting that the

primitives of perception are complex and multidimensioned, there is within his mode of analysis a suggestion of a different volatility, of a certain polymorphous quality. I shall take one particularly ambiguous example, that of three-dimensional visual effects from two-dimensional drawings (long favored examples for psychology experiments). Merleau-Ponty contends:

> Organization in depth is destroyed if I add to the ambiguous drawing not simply any lines (Fig. 3 stubbornly remains a cube) but lines which disunite the elements of one and the same plane and join up those of different planes (Fig. 1). What do we mean when we say that these lines themselves bring about the destruction of depth? Are we not taking the language of associationism? We do not mean that the line EH (Fig. 1), acting as a cause, disorganizes the cube into which it is introduced, but that it induces a general grasp which is not the grasp in depth.

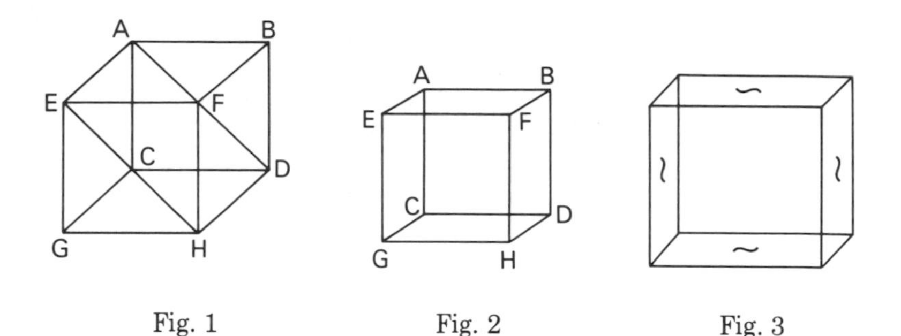

Fig. 1 Fig. 2 Fig. 3

> It is understood that the line EH itself possesses an individuality only if I grasp it in that light, if I run over it and trace it out myself. But this grasp and this delineation are not arbitrary. They are indicated or recommended by phenomena. The demand here is not an overriding one, simply because it is a question of an ambiguous figure, but in a normal visual field, the segregation of planes and outlines is irresistible; for example, when I walk along an avenue, I cannot bring myself to see the spaces between the trees as things and the trees themselves as a background. It is certainly I who have the experience of the landscape, but in this experience I am conscious of taking up a factual situation, of bringing together a significance dispersed among phenomena, and of saying what they of their own accord mean.[43]

What may be noted from this discussion is that: (a) such ambiguous phenomena, in his interpretation, have a certain gradation of ambiguity but are clearly polymorphous in some degree; and (b) the gradation of polymorphy, however, is related to certain "bodily engagement" possibilities. Thus Merleau-Ponty contends that reversibilities from such abstract and disengaged drawings are not repeatable within the more engaged praxis of motion (walking down the tree-lined avenue). Yet, within the ambiguous gestalt that the figures present, there is a range of perceptual polymorphy.

From this limited exposition of the praxis-perception model emerging from the *Phenomenology of Perception* it should be clear that it focuses upon what I call microperception, that is, the action and perception which occurs in our bodily or incarnate engagement with the immediate environment or world. What remains unclear is how this model of bodily-perceptual engagement relates to the development of either the new philosophy of science or the philosophy of technology.

I will make only one basic comment with respect to the philosophy of science. By adding an analysis of perception and indicating how it is a constant in the emergence of truth or rationality, Merleau-Ponty clearly belongs to that group of thinkers who would necessarily see the praxis of science as both an engaged and an embodied form of action. It should also be obvious, particularly from the central role gestalt models play, that the analysis of perception supplements what Kuhn later developed in *The Structure of Scientific Revolutions.* Indeed, the specific analysis of perception is considerably beyond the depth of either Kuhn's or Wittgenstein's use of perception. But this is not to claim that Merleau-Ponty applied his insights to the institution of science. That was not his aim.

If the role of perception, especially at the microperceptual order of things, does not immediately seem to fit the concerns of philosophy of science, the opposite is the case with respect to the philosophy of technology. The materiality of body carries with it, particularly with regard to artifact use, certain rather immediate implications. Merleau-Ponty offers three examples in the discussion of bodily spatiality:

> A woman may, without any calculation, keep a safe distance between the feather in her hat and things which might break it off. She feels where the feather is just as we feel where our hand is. If I am in the habit of driving a car, I enter a narrow opening and see that I can "get through" without comparing the width of the opening with that of the wings, just as I go through a doorway without checking the width of the doorway against that of my body.[44]

In Polanyi's parlance, these phenomena would be examples of *tacit knowledge,* since they are a kind of "know how" without explicit conceptual judgment attached to them. But they are more—they are examples of how artifacts (technologies) may be used or experienced in use. They are examples of what I call *embodiment relations.* Such relations are existential (bodily-sensory), but they implicate how we utilize technologies and how such use transforms what it is we experience *through* such technologies. Merleau-Ponty is more precise in his example of the blind man's cane:

> The blind man's tool has ceased to be an object for him, and is no longer perceived for itself; its point has become an area of sensitivity, extending the scope and active radius of touch, and providing a parallel to sight. In the exploration of things, the length of the stick does not enter expressly as a middle term: The blind man is rather aware of it through the position of objects than the position of objects through it. The position of things is immediately given through the extent of the reach which carries him to it, which comprises besides the arm's own reach the stick's range of action.[45]

I contend that this experiential phenomenon has implications not only for the philosophy of technology, but for the philosophy of science. The feather, the automobile, and the cane fall into the same existential use as many scientific instruments, a use that has simply been ignored in most of the standard analyses but which belongs essentially to the expansion of insight needed within a new philosophy of science.

If the primary contribution toward both philosophy of science and, potentially, the philosophy of technology may be found in the analysis of microperception in its sensory-bodily dimensions, there is also a growing awareness on Merleau-Ponty's part of what I call macroperception in its interaction with microperception. Unfortunately, these later insights received were far less developed, but they are indicated in his later work, *The Visible and the Invisible.* He was clearly aware of the role of culture in relation to perception:

> It is a remarkable fact that the uninstructed have no awareness of perspective [in art] and that it took a long time and much reflection for men to become aware of the perspectival deformation of objects. . . . I say that the Renaissance perspective is a cultural fact, that perception itself is polymorphic and that if it became Euclidian [sic], this is because it allows itself to be oriented by the system. . . . What I maintain is that: there is an informing of perception by culture which enables us to say that culture is perceived.[46]

Here is a clear, emerging awareness of the interaction between and the interrelatedness of micro- and macroperception.

Although there does not seem to be any indication that Merleau-Ponty applied the gestalt and phenomenological notion of field and focus or figure and ground to this interrelation, his analysis is clearly open to that suggestion. But Merleau-Ponty does not simply absorb microperception into macroperception in a Wittgensteinian nor, as will follow, a Foucaultean way.

The late Michel Foucault (d. 1983) will be the last thinker I shall discuss as a European counterpart of the new philosophy of science.

Such a choice might seem strange to some phenomenologists, because Foucault characterized himself as antiphenomenological. Yet in spite of this self-characterization I shall argue that Foucault develops precisely the praxis-perception model which is needed for a philosophy of technology. He was, in fact, a student of Merleau-Ponty; there is substantial internal evidence that much of his work was effectively a response to Merleau-Ponty. (He was an avid reader of Heidegger as well.)

On the other side, there is an as yet unappreciated appropriateness to placing Foucault in this discussion. He was a strict contemporary of Thomas Kuhn; and his first major work, *Madness and Civilization,* was published in 1961, actually preceding Kuhn's *Structure of Scientific Revolutions* by one year. And while there has been something of a fad concerning Foucault in certain circles—particularly among postmodernists and literary critics—it is only very recently that his impact on philosophy of science has begun to emerge. There are several reasons for this. First, his taste for such phenomena as madness, punishment and sexuality—his "archeologies," as he calls them—is less likely to appeal to the sparse soberness of much of the philosophy of science establishment. Second, given his notion of discourse and power contexts and the apparent application primarily to the human sciences, it may have seemed that Foucault was at best tangential to the philosophy of science. That has changed in the last few years, thanks in part to the influence of Hubert Dreyfus, who with a whole generation of recent bilingual younger philosophers of science has succeeded in linking Heidegger and Foucault in a new form of discourse/power context.

I shall be taking a quite different tack by looking at Foucault through the praxis-perception model of interpretation I am now building. Thus, part of the task is to re-locate Foucault in that interpretation. In a strikingly parallel way, his *Order of Things* (1966) is a reinterpretation of the human sciences, as Kuhn's *Structure of Scientific Revolutions* was of the natural sciences. Foucault's archeologies and histories of perception also emphasize sudden and discontinuous shifts in perspective and, like Kuhn, exemplify shifts of the macroperceptual domain. Foucault is thus a pivotal figure in this juncture between phenomenology and the new philosophy of science.

He most explicitly defended his method in *The Archeology of Knowledge* (1969), in which he made his stylistically flamboyant and deliberately contrary strategy most apparent. He argued that if standard intellectual history is linear, his would be nonlinear; if the standard was continuous and anticipatory or evolutionary, his would be discontinuous and anti-anachronistic. The same contrary strategy applied to any history-through-individuals approach, for which he substituted his version of *annales* history. The latter focuses upon anonymous history, even "unconscious" history, which belongs to no one in particular but does belong to *social practice.*

However, such seemingly contrary perversity, can also be seen as a kind of deliberate gestalt switching. What was for Kuhn a kind of occurrence, in Foucault's practice becomes a deliberate device. This practice, I would argue, is familiar to anyone steeped in phenomenology. It is the deliberate use of variational method which traces all the way back to Husserl, and this use is clearly visible in Foucault's own praxis.

However, the results of Foucault's archeologies and histories of perception run so parallel to the phenomena which Kuhn calls paradigm shifts that one wonders whether or not Foucault may not be right regarding the appearance of virtually simultaneous new practices. (This is not to deny that Kuhn's interpretation of paradigm shifts remains at the level of a more conceptualistic shift than those illustrated by Foucault, which presumably evidence a discourse-praxis.)

I have suggested that Foucault belongs within the overall phenomenological trajectory despite his explicit self-denials. Not only was he a student of Merleau-Ponty, but the outline of *Order of Things* betrays simply too many specific responses to Merleau-Ponty's writings to go unnoticed. I shall note a few of these to make the point: Merleau-Ponty's favored examples came primarily from aesthetics, and at one point he made the claim that whereas one could have language about language, one could not have painting about painting. *Order* opens with a discussion of Velasquez's *Las "Meninas,"* a painting which shows a painter painting, but which more deeply is a mirror-phenomenon of reflections, a painting about painting. The very next chapter is "The Prose of the World," a title taken directly from one of Merleau-Ponty's later collections. So just as the subject of Velasquez's painting remains invisible within the representation, so does Merleau-Ponty remain unseen in *Order.* Contrary to his statements but based on his own work, I shall then place Foucault within the trajectory of a phenomenological preoccupation with perception and praxis. His negative response to phenomenology remains at a surface level, but the model of interpretation which he follows, while uniquely modifying the tradition, remains recognizably within it.

The modifications which Foucault enacts within a phenomenological trajectory can be seen as following some of the inner, but often undiscerned, implications of this tradition. Husserl, particularly in his early writings, retained the traditions of the transcendental philosophies, which run from Descartes through Kant. These philosophies located the source and structure of knowledge within a subject. It is this "phenomenology" which Foucault strongly rejects:

> If there is one approach that I do reject, however, it is that (one might call it broadly speaking, the phenomenological approach) which gives absolute priority to the observing subject, which attributes a constituent role to the act, which places its own point of view at the origin of all historicity—which, in short, leads to a transcendental consciousness.[47]

What Foucault is rejecting is both the transcendental and foundational framework within which Husserl remained. But the rejection of both these aspects of an older tradition already had been discarded by Merleau-Ponty and Heidegger—dramatically so in their later works. Post-Husserlian phenomenology was already existential and hermeneutic in form. And it is from this unstated base that Foucault really begins.

The shift Foucault takes combines certain features of structuralism and semiotics—also powerful French movements—into a still basically phenomenological aim. He rejects the subjectivist aspects of phenomenology regarding a knowing subject and replaces this with a framework focusing upon discourse-praxis and social or macroperception. "It seems to me that the historical analysis of scientific discourse should, in the last resort, be subject, not to a theory of the knowing subject, but rather to a theory of discursive practice."[48]This places praxis and perception at the linguistic and cultural level.

In his initially European context, Foucault's critique belongs to a whole movement of iconoclasts who have attacked the older transcendental and foundational philosophies. The "self" is to be decentered, "intentionality" interpreted as deliberate acts of knowledge overthrown, and the role of knowing consciousness displaced. In their stead, one is to examine anonymous histories, "unconscious" models of knowledge, and the discontinuities within perceptual histories. His most notorious notion, usually taken out of context, is his frequently quoted claim that "Man" is a recent invention:

> Strangely enough, man—the study of whom is supposed by the naïve to be the oldest investigation since Socrates—is probably no more than a kind of rift in the order of things, or, in any case, a configuration whose outlines are determined by the new position he has so recently taken up in the field of knowledge. Whence all the chimeras of the new humanisms, all the facile solutions of an "anthropology" understood as a universal reflection on man, half-empirical, half-philosophical. It is comforting, however, and a source of profound relief to think that man is only a recent invention, a figure not yet two centuries old, a new wrinkle in our knowledge, and that he will disappear again as soon as that knowledge has discovered a new form.[49]

One can immediately see why such an apparently outrageous claim drew attention. But its sense may be revealed only by taking account of the substitutions Foucault makes in the praxis-perception paradigm.

Briefly put, the history of the *perception* concept I have followed here is rooted in the late Husserl, where perception in its micro- or sensory connotation does remain within the overall traditions of a transcendental and *foundational* philosophy. The grounding perception of the lifeworld is both universal and foundational—at least, in its basic outline. Moreover, it is a perception which is *multidimensional* in that its aim is at the secondary kind of familiarity or cultural taken-for-grantedness, is already a modification upon the presumed foundation. The Galilean revolution is one which can be culturally "perceived" and taken-for-granted over centuries as the normative perspective of the physical sciences.

In existentializing Husserl's entire notion of the perceptual arc, Merleau-Ponty moves ever farther away from both transcendental and foundational interpretations. While the living body provides a weighted focus, a located point of view which is also the locus of action, it is clearly *not* the foundation of the world or even of knowledge.

In Merleau-Ponty, intentionality had already become existential, and in many of its aspects, tacit or not explicitly "conscious." The body, while remaining concrete and material, was not the "universal" foundation of knowledge, although it was the means of access to the world. But the world was seen to be both ambiguous and polymorphous within multidimensional perception. In the later works, perception itself relates to cultural contexts, for if different historical eras can discern depth in paintings differently, perception itself must exhibit a polymorphic quality.

Putting the phenomenological history of perception in this way, one can see that Foucault continues its radicalization. In turning to "body," the movement is one which makes body highly ambiguous and polymorphic. The "body" is not simply a material body but also a cultural body (as in the "body politic"). *Discipline and Punish* (1975) begins with a gruesome description of the drawing and quartering of a regicide in 1757. The condemned's body was here the focal point of an entire political praxis, each step of the cruel and drawn-out execution virtually ritualized as the power of the sovereign is concretely applied to the body of the prisoner. The entire act occurs within sight of the populace so that the power of the king may be seen.

In a paradigm shift with respect to punishment, Foucault then contends that a more modern form occurs as a massive set of substitutions. "The body as the major target of penal repression disappeared. Punishment . . . [becomes] the most hidden part of the penal process."[50] But this shift from public to hidden executions, and then to long imprisonment out of public sight, also contains a shift in the entire gestalt. Its focus is much more upon the control of the inner person and the removal from the new form of the "body politic." "When the moment of execution approaches, the patients are injected with tranquilizers. A utopia of judicial reticence: Take away life, but prevent the patient from feeling it; deprive the prisoner of all rights, but do not inflict pain. . . . "[51]

In our context, such dramatic paradigm shifts in the order of control are modifications upon the praxis-perception model toward an analysis of structured macroperceptions. Here again, Foucault's actual analysis, even though interested in the phenomenon of power, follows a parallel with Kuhn.

He is close to that perspective for another reason. Foucault took much of his modification upon the praxis-perception model from structuralism and its linguistics (de Saussure and the traditions of French semiotics). Interestingly, although formed from very different sources, this background resulted in an essentially Wittgensteinian stance in which perception is always located within language contexts. Perception belongs inextricably to language. Language or discourse, in Foucault, locates and contexts perception. The result of these moves is the trademark of the new philosophy of science, a subtle attentiveness to discontinuities, particularly in the scientific perception of the world.

The Order of Things (1966) comes closest to philosophy of science concerns. The preface begins with a "Chinese encyclopedia":

> This book first arose out of a passage in Borges, out of the laughter that shattered, as I read the passage, all the familiar landmarks of my thought—our thought, the thought that bears the stamp of our age and our geography—breaking up all the ordered surfaces and all the planes with which we are accustomed to tame the wild profusion of existing things, and continuing long afterwards to disturb and threaten with collapse our age-old distinction between the Same and the Other. This passage quotes a "certain Chinese encyclopaedia" in which it is written that "animals are divided into: (a) belonging to the Emperor, (b) embalmed, (c) tame, (d) sucking pigs, (e) sirens, (f) fabulous, (g) stray dogs, (h) included in the present classification, (i) frenzied, (j) innumerable, (k) drawn with a very fine camelhair brush, (l) et cetera, (m) having just broken the water pitcher, (n) that from a long way off look like flies." In the wonderment of this taxonomy, the thing we apprehend in one great leap, the thing that, by means of the fable, is demonstrated as the exotic charm of another system of thought, is the limitation of our own, the stark impossibility of thinking that.[52]

One can immediately sense that something strange is afoot. But in spite of the extreme example, Foucault's thesis itself is more modest. He traces the distinctive features of what he calls *epistemes,* periods of the order of knowledge, and shows how they are both particular organizations of knowledge and perception and how they are radically discontinuous with one another. They are paradigm shifts within our history. The first such shift is virtually canonical, from the Middle Ages to the time of the rise of science through the Renaissance, examples of which are vividly analyzed in *Order*. But Foucault sees more: He argues that there have been radical shifts in the order of knowledge between the Renaissance and the Classical periods (the seventeenth to eighteenth centuries, usually referred to in philosophy as the modern period) and again between the nineteenth and present centuries. In short, where the dominant traditions of intellectual history have assumed continuity, Foucault discerns discontinuity as evidenced by his perceptual history. I shall examine briefly a few samples of this analysis to show not only how Foucault clearly parallels the thrust of the new philosophy of science but subtly advances its insights through the examination of such different gestalts. I shall restrict my examples to those closest to sciences. In this case, the fields will be botany and biology.

The question in this context is how plant and animal life is perceived. In the broad sense employed by Foucault, a perception is an ordered knowledge and is revealed in the way the order is shaped, the way it is represented and described, and the way observation-perception occurs.

Foucault's characterization of the sixteenth-century *episteme* would most likely be the easiest to recognize. The organizing theme which characterized this era was the great, complex web of *resemblance.* "It was resemblance that largely guided exegesis and the interpretation of texts; it was resemblance which organized the play of symbols; made possible knowledge of things visible and invisible, and controlled the art representing them."[53] Moreover, the perceiving of this vast, complex but tightly interwoven Medieval world was simultaneously a perceiving and a *reading.* Foucault argues that in this period, reading is the very model of language; nature, the heavens, all things are to be read as though a text. It was the age of *interpretation.*

I shall not attempt to deal with the subtlety or the complexity of his analysis of resemblance, marked by a number of internal relations which linked the era, or *episteme,* into a coherent paradigm. But I shall relate this to the *perception* of the variable botanic and biological life which takes its place within this episteme. To observe or perceive a plant or animal is to do so within the context of the order of things Medieval. Quoting a text of the period, Foucault notes, "Is it not true that all herbs, plants, trees and other things issuing from the bowels of the earth are so many magic books and signs?"[54]

Such a paradigm is not limited to botany or biology. It is a well known fact that this mode of knowledge is also a key to the organization of the great cultural monuments of the time—the European cathedrals. Here one "reads" the history of the world in glass and stone, and the whole is organized into its balance and overlay of similitudes. And if the primary text, the Bible, does not in itself provide the balance needed, extra-biblical examples are added to complete the picture.

Symmetry, resemblance, and the great Medieval notion of analogy context the very perception of botany.

> There exists a sympathy between aconite and our eyes. This unexpected affinity would remain in obscurity if there were not some signature on the plant, some mark, some work as it were, telling us it is good for diseases of the eye. This sign is clearly legible in its seeds! They are tiny dark globes set in white skinlike coverings whose appearance is much like that of eyelids covering an eye. It is the same with the affinity of the walnut and the human head! What cures "wounds of the pericranium" is the thick green rind covering the bones—the shell of the fruit; but internal head ailments may be prevented by the use of the nut itself "which is exactly like the brain in appearance."[55]

Such an organization of knowledge—and particularly of perception—must seem as strange today as the "Chinese encyclopedia" which startles Foucault.

This strangeness may be revealed in another way: What if we were to take seriously the following example?

> The world is simply the universal "convenience" [close relation or similitude] of things; there are the same number of fishes in the water as there are animals, or objects produced by the nature of man, on the land (are not fishes called *Episcopus,* others called *Catena,* and others called *Priapus?*); the same number of beings in the water and on the surface of the earth as there are in the sky, the inhabitants of the former corresponding with those of the latter. . . . [56]

Were we to reverse the usual negation by the present view upon this example—which in the textbooks would probably take the shape of criticizing the non-empirical sense of organization—and ask, do we *know* this to be false? we would immediately see that not only do we not know it to be false but we do not even have the means to determine its truth or falseness. Whole earth measurements by means of satellites and whole population demographies of plant and animal life simply do not exist, even in the twentieth century.

Rather, the very question strikes us as strange, since we simply no longer understand or perceive things under this paradigm. What could it possibly mean if the numbers *were* right? This is the point Foucault makes. Resemblance, as organization of things, simply drops out by the seventeenth century.

> At the beginning of the seventeenth century, during the period that has been termed, rightly or wrongly, the Baroque, thought ceases to move in

> the element of resemblance. Similitude is no longer the form of knowledge, but rather the occasion of error, the danger to which one exposes oneself when one does not examine the obscure region of confusions.[57]

Here is a radical paradigm shift, one which is extreme enough to invert the very role of resemblance. It is a discontinuity in cultural perception:

> Discontinuity—the fact that within the space of a few years a culture sometimes ceases to think as it had been thinking up till then and begins to think other things in a new way—probably begins from outside, from that space which is, for thought, on the other side, but in which it has never ceased to think from the very beginning.[58]

Here, independently parallel to Kuhn, is a European example of discontinuities rooted in a radical shift of the organization of knowledge and the rearrangement of perception. Moreover, this is a gestalt shift, not of one variable but of an entire constellation. Within the Medieval *episteme* there is "a nondistinction between what is seen and what is read, between observation and relation, which results in the constitution of a single, unbroken surface in which observation and language intersect to infinity."[59] Thus, with respect to the bestiaries and the fabulous botanies, it is not so much the non-empirical character which differentiates the descriptions found there as the very concept of what counts in a perception. The individual plant or animal simply belongs, in its very visibility, to the wider invisibility of relations, which include the multiple dimensions ranging from the heavens to hell. Observation, if it may be called that at all, must include not only colors, which themselves relate outward, but their relations to the whole: "The bright colors of the flowers reproduce, without violence, the pure form of the sky."[60] These relations include its history, its uses, its theological connections to the virtual infinity of the Medieval *episteme.*

In his reinterpretation of the accepted paradigm shift between Medieval and Modern periods, Foucault characterizes this shift as a gestalt. It includes both differing perceptions and a different understanding of language itself: "The profound kinship of language with the world was thus dissolved. The primacy of the written word went into abeyance. And that uniform layer, in which the *seen* and the *read,* the visible and the expressible, were endlessly interwoven, vanished too."[61]The same applies to the understanding of perception. It is severed from its relations and histories. "The eye was thenceforth destined to see and only to see, the ear to hear."[62] This was also the time when the senses began to be divided into discrete dimensions. There was a shift in the role and even the understanding of the senses.

Neither Foucault nor Kuhn is able to answer why such shifts occur.

Foucault's suggestion that erosion occurs from the outside parallels Kuhn's sense of internal accumulated anomalies. But in both cases it becomes clear that the arguments which establish the new paradigm as ascendent become possible and plausible only after such a shift, a different way of seeing, has occurred. Thus, once resemblance changed its value, there could emerge a wide spectrum of arguments which attack it. Bacon's attack upon idols, Descartes's attack upon resemblance, and most scathingly, Hume's attacks upon analogy are made from an already established counterview. And Foucault sees, rightly, I think, that the seventeenth century arguments were far more than a call for empirical, testable experiments; their basic thrust was against knowledge by resemblance. Our penchant to read the attacks as anticipations of Positivism is itself anachronistic.

To this point we are on at least partly familiar ground. The striking distance between us and the Medievals is too great not to notice. The Medieval *episteme* is at best a monument—interesting, impressive, imaginative, but archaic. But not so the Modern period, with its association to the rise of Modern Science. This period we more often take as contiguous with ours, even within some of the dominant traditions of contemporary philosophy. Foucault argues to the contrary, and in effect, claims that the organization of Modern knowledge and its version of perception *is as archaic and distinct from the contemporary era as other no-longer-utilized paradigms.* Our reading of the Modern era, Foucault contends, has become naïvely anachronistic because the modern way of ordering knowledge has been completely forgotten or missed.[63]

The first set of substitutions is at least partly familiar. For resemblance—which now becomes the domain of confusion to be rejected and replaced—the Modern era established a notion of *Order,* whose method of organization is to be geometrical *analysis.* "This relation to *Order* is as essential to the Classical (Modern) age as the relation to *Interpretation* was to the Renaissance."[64] Indeed, it is more the sense of order than of measurement which takes precedence early in the period:

> . . . Relations between beings are indeed to be conceived in the form of order and measurement, but with this fundamental imbalance, that it is always possible to reduce problems of measurement to problems of order. So that the relation of all knowledge to the mathesis is posited as the possibility of establishing an ordered succession between things, even nonmeasurable ones. In this sense, *analysis* was very quickly to acquire the value of a universal method.[65]

We are obviously close to the same observation made by Husserl. But by his use of disciplines other than geometry and physics, a different perspective is gained. For once beyond the close linking of shapes and spatial arrangements possible in the first two sciences, once the organic world is entered, other problems arise, particularly for Galilean and Cartesian "mechanism." In the realm of botany and biology, "The centre of knowledge, in the seventeenth and eighteenth centuries, is the *table.*" Arguments occur within the context of a table. Where are things to be located with respect to the table? *Natural history* is born, and its order is *visible* taxonomy.

The changes which occur are simultaneous, like the reversal of a figure and ground. But one can note this by looking at the variables of meaning—for example, "history." Foucault asserts that "In the sixteenth century, and right up to the middle of the seventeenth, all that existed was histories."[66] But by 1657, with the publication of Jonston's *Natural History of the Quadrupeds,* there is a shift such that what had been "history" to the Medievals simply drops out. "Jonston subdivides his chapter on the horse under twelve headings: name, anatomical parts, habitat, ages, generation, voice, movements, sympathy and antipathy, uses, medicinal uses."[67]

Here we still have the sense of the Chinese encyclopedia compared to the present, but the being's relations are nevertheless changed.

> The essential difference [between Jonston and his predecessors] lies in what is *missing* in Jonston. The whole of animal semantics has disappeared, like a dead and useless limb. The words that have been interwoved in the very being of the beast have been unravelled and removed: and the living being, in its anatomy, its form, its habits, its birth and death, appears as though stripped naked.[68]

In this sense, the natural history which is born within the context of the table of order *reduces* the multiple relations and dimensions of the beings from their previous organization.

What was both a history, yet also a character of the animal is then what is displaced and then disappears by means of substitutions. Linnaeus prescribes a specific plan of description which must follow an order: name, theory, kind, species, attributes, use—and then, as a kind of useless appendix, *litteria.*[69] In this process of reduction, history changes meaning. "The Classical [Modern] age gives history a quite different meaning: that of undertaking a meticulous examination of things themselves for the first time, and then of transcribing what it has gathered in smooth, neutralized, and faithful words."[70]

This organization even differs, Foucault argues, from that of the Renaissance, which he views as a transition between the Medieval and the Modern eras. The Renaissance also collected and was interested in exotic plants and animals, but more as theater. "What came surreptitiously into being between the age of the theatre and that of the catalogue was not the desire for knowledge, but a new way of connecting things both to the eye and to discourse. A new way of making history."[71]

If the meaning of what counts as history changes, so does the variable of *perception*. Perception is linked in a different way to language: "Natural history is nothing more than the *nomination of the visible*. [Italics mine]"[72] Here is a new way of seeing: It is the transformation of previous perception into "observation." In Foucaultean terminology, one can say observation is born:

> Observation, from the seventeenth century onward was a perceptible knowledge furnished with a series of systematically negative conditions. Hearsay is excluded, that goes without saying but so are taste and smell, because their lack of certainty and their variability render impossible any analysis into distinct elements that could be universally acceptable. The sense of touch is very narrowly limited to the designation of a few fairly evident distinctions (such as that between smooth and rough); which leaves sight with an almost exclusive privilege, being the sense by which we perceive extent and establish proof, and in consequence, the means to an analysis *partes extra partes* acceptable to everyone.[73]

The new Order now determines *which* senses are to be favored and it also determines how these senses are to be focused. That which is clear and distinct and spatially extended rises in importance over other possibilities. In short, perception becomes weighted in favor of vision, is virtually reduced *to* vision; but more, it also selects that which, within the visible, is to count, a reduction *of* vision. This is so much the case that the metaphysics of primary and secondary qualities even enters into what can count within vision. "The area of visibility in which observation is able to assume its powers is thus only what is left after these exclusions: a visibility freed from all other sensory burdens and restricted, moveover, to black and white."[74] Finally, this vision is structured according to geometrical values with emphasis upon lines, surfaces, forms, and reliefs.[75]

In passing, I wish to note that it is in this context that Foucault makes one of his rare comments about what is here termed *technology*. It is clear that for Foucault it is the order of things which determines the selection and use of technologies and not the other way around. "In fact, it was the same complex of negative conditions that limited the realm of experience and made the use of optical instruments possible."[76] In this respect, Foucault, not unlike the other new philosophers of science, remains Platonistic.

Observation is born in the seventeenth century and is a particular ordered perception, linked to the visible but also within the visible, strictly and selectively ordered. "To observe, then, is to be content with seeing—with seeing a few things systematically."[77] It is to leave out precisely what Husserl called the plenary qualities of things. Science becomes the immense task of naming and organizing the world of the visible into the classifications and tables of the Modern era. Botany and biology become natural histories (classifications) and taxonomies. And the arguments internal to this Order are those about how many, where, and how organized the things must be—this is the normal science of the seventeenth and eighteenth centuries.

Foucault argues that this way of seeing the world has in effect disappeared. Nor should it be anachronistically understood as some kind of anticipation of the next epistemic organization. It precedes the episteme of the nineteenth century, to be sure; but with the invention of biology proper (Foucault refuses to apply this term to the study of animal life prior to the nineteenth century), there occurs yet another radical paradigm shift. And with this shift appears another gestalt arrangement which reorganizes all the variables simultaneously.

Again restricting my examples to botany and to biology as characterized by Foucault, there is both a change in the order of knowledge and in the realm of perception. From placing things upon a table arranged according to a set of variables open to vision, the invention of biology shifts to that which is *internal* and *invisible.* Its paradigmatic science is no longer a natural history, but comparative anatomy. "From Cuvier onward, it is life in its non-perceptible, purely functional aspect that provides the basis for the exterior possibility of a classification."[78]

The new episteme revalues the plenum in a new way, with one dominant substitution, that of the interior for the exterior. "The first of these techniques is constituted by comparative anatomy: This discipline gives rise to an interior space, bounded on the one side by the superficial stratum of teguments and shells and on the other by the quasi-visibility of that which is infinitely small."[79] There is, then, a kind of inversion in which what had been important—precisely, the exterior of shapes and number and arrangement, etc.—is now taken as secondary to the interior microstructure investigated by anatomy.

While my focus here remains upon Foucault's interpretation of the paradigm shift from the eighteenth to the nineteenth centuries, I would first enter what will become a major disagreement with its characterization. By terming the invention of biology as a shift to the "invisible," there is a shift to what had been either not attended to or attended to in a simple, manifest way. Biology does not shift to the *invisible,* but to a different kind of visibility which Foucault terms quasi-visibility, the internal and the microstructured. But this domain is precisely what the technology—instrumentation—of science made available to the nineteenth century as never before. The new seeing is by means of instrumentation, a not inconsequential phenomenon! It was stimulated in part by the invention of staining techniques, which made visible what was previously invisible even under the microscope.

Because I should by now have established that Foucault ascribes to the perspective of the new philosophy of science in which the different eras of science are established by reorganizations of *seeing as,* I shall not go into as great detail on the eighteenth- and nineteenth-century *epistemes,* as in the previous examples. As with any radical discontinuity, I would simply point out that the reorganization of the order of things includes the whole language-practice-perception gestalt.

In characterizing the discontinuity, it is interesting to note that just as large sectors and relations between things drop out between the sixteenth and seventeenth centuries, there are vast eliminations and substitutions between the Modern and the nineteenth-century life sciences. Foucault notes that these include such elements as "The parallelism between classification and nomenclature is thus . . . dissolved,"[80] and "To classify . . . will no longer mean to refer the visible back to itself."[81] Even the notion of Nature as conceived by natural histories radically changes:

> "Nature," too, disappears . . . as the homogeneous space of orderable identities and differences. This space has now been dissociated and as it were opened up in depth. Instead of a unitary field of visibility and order, whose elements have a distinctive value in relation to each other, we have series of oppositions, of which the two terms are never on the same level.[82]

Positive substitutions are those revealed by anatomies: functions, the similarities of organic processes which are identical in spite of radically different arrangements of bodily parts, a hierarchy of organs related to the different functions, etc.

With the new reemergence of the invisible, Foucault notes that something like a revival of "Medieval" relationality occurs. A new system of configurations and signs pointing beyond themselves arises in the anatomical, functional sciences of biology.

> The . . . propulsive limbs belong to the plan of the organic structure, but only as a secondary character! They are therefore never eliminated, or absent or replaced, but they are "masked sometimes as in the wings of a bat and the posterior fins of seals"; it may even happen that they are "denatured" by use as in the pectoral fins of the cetaceans. . . . You perceive that there is always a sort of constancy in the secondary characters in accordance with their disguise.[83]

Such occurs with the substitution of biology for natural history. It is only by means of archeological foreshortening, Foucault argues, that we fail to see how thoroughly the Modern paradigm has disappeared. It is as archaic in the context of the contemporary as the Medieval order of knowledge was to it.

By now the parallelism of Foucault with Kuhn, without discounting either, since their interpretations were strictly contemporaneous, should be established. I have not accentuated what could be equally important distinctions because the purpose here is to solidify what I take to be a certain underlying development of a *praxis-perception* model of interpretation which the new philosophy of science directs toward its subject matter. And equally, I have tried to demonstrate that this model has been rather deeply ingrained in the Euro-American phenomenological traditions, with or without the recognition of the dominant philosophies in North America.

This has been a look at a certain *moment* in the rise of what I have called New philosophies of science. And although I have suggested that there is an overlooked, forgotten background element which pervades even the New philosophy of science—science's technologies—neither Kuhn nor his European counterparts advance that issue. Also, if Kuhn and his European counterparts place science in a praxical context, even one marked by an appreciation of its perceptual dimensions, they still do not make the link to the embodiment of science which historically transformed Modern Science as a crucial element in either praxis or perception.

Guilty of this same oversight are those few philosophers who have begun to be interested in philosophy of technology out of the older strands of the philosophy of science. I have already noted Bunge's attempt to force what he calls technophilosophy into such a mode. For Bunge, technology is to be placed under a nomological model of interpretation. Thus, none of the predecessors put forth nor focus upon the more concrete aspects of human-technology *praxis* or *perception.* The time has come to shift more directly to what might well be the roots of a philosophy of technology, in contrast with—but also in relation to—the new philosophy of science.

III. Philosophy of Technology

The new philosophy of science in both its Anglo-American and European contexts has represented a change in sensibility and in perspective. It is directed away from what may be called a statics of conceptual and logical relations (the nomological model) toward what may be called a dynamics of "seeing" (the potential praxis-perceptual model). But if this is so, there remains a certain vestigial Platonism throughout the new philosophy of science. It remains insensitive to the material embodiments of science, to its technological dimensions.

One clear and crucial domain in which this neglect is obvious is with respect to *instrumentation*. In contrast to its ancient antecedents, contemporary science is clearly technologically embodied. Instruments form the conditions for and are the mediators of much, if not all, current scientific knowledge. They are the concrete and material operators within scientific praxis. Yet little has been done, even by the new philosophers of science, regarding the effect instruments play upon the paradigms or epistemes which occur in science. I shall use this variable to make the transition to a possible philosophy of technology.

It is not the case that instruments are totally overlooked within the new philosophy of science, but they clearly play a background role. A brief review of the four figures just discussed will illustrate this lack of development: In Kuhn's case, there is an awareness that instruments play a role in observation. "Herschel, when he first observed the same object twelve years later . . . *with a much improved telescope* of his own manufacture . . . was able to notice an apparent disk-size that was at least unusual for the stars."[1]

This observation played a role in the shift of star/planet interpretation, but Kuhn seems to simply assume the instrument, even though it plays here the role of condition of the observation. Kuhn only hints at what he calls an instrumental expectation.

> In short, consciously or not, the decision to employ a particular piece of apparatus and to use it in a particular way carries an assumption that only certain sorts of circumstances will arise. There are instrumental as well as

> theoretical expectations, and they have played a decisive role in scientific development.[2]

Only rarely does Kuhn recognize the crucial role a new piece of apparatus may play as the *prelude* to a paradigm shift. One example is the development of the Leyden jar, which in effect opened the way to the discovery of electrical phenomena.[3]

If we turn from Kuhn to his European counterparts, the situation is not markedly improved. There is some recognition in Husserl, though rarely followed through, of the role of technologies in scientific praxis. In part, this is noted because at the base level, lifeworld perception is related to material entities. But given the interest in the accumulation and progress of idealities, this material dimension plays a facilitation role. One interesting example is the clearly crucial development of the technologies of language, or *writing:* "The important function of written, documenting linguistic expression is that it makes communications possible without immediate or mediate personal address; it is, so to speak, communication become virtual. Through this, the communalization of man is lifted to a new level."[4]

Recalling Husserl's carpentry or furniture example, we have already noted that praxis involves material processes which potentially hold trajectories (the preference for and following of straight lines, smooth surfaces, etc.). Yet, given Husserl's focus upon the acquisition of a pure geometry, the role of material, praxical activity remains, at most, a material condition for this acquisition:

> The empirical art of measuring and its empirically, practically objectivizing function, through a change from the practical to the theoretical interest, was idealized and thus turned into the pure geometrical way of thinking. The art of measuring thus became the trailblazer of the ultimately universal geometry and its "world" of pure limit-shapes.[5]

In short, technologies are conditions for or occasions of advance in idealization.

Of course, Merleau-Ponty's interests simply do not lend themselves immediately to what I am calling a philosophy of technology, which focuses upon the material conditions and mediations relating to knowledge gathering in general or to science in particular. Yet his analysis of one set of artifacts in bodily self-awareness is highly suggestive. The woman whose feathered hat becomes intuitively *embodied* in her motion within an environment, the tacit bodily knowledge which the automobile driver demonstrates with his "instrument" of motion, and the highly skilled, learned motile knowledge of the blind person with the cane all demonstrate a phenomenological dimension to instrumentally mediated knowledge.

Yet it remains that this analysis of the mediation of perception through technologies is not developed by Merleau-Ponty as an aspect of the philosophy of science.

Foucault actually returns to a much older prejudice by not placing the instrument solely in the role of a mere application of science, but he also seems to disclaim the role of instrumentation as crucial to certain developments of scientific praxis. Foucault holds that there was both a reduction *to* visibility and a reduction *of* visibility which characterized scientific praxis at the outset of the Modern era:

> Observation, from the seventeenth century onward, is a perceptible knowledge furnished with a series of systematically negative conditions. Hearsay is excluded . . . but so are taste and smell, because their lack of certainty and their variability render impossible any analysis into distinct elements that could be universally acceptable.[6]

This simultaneous reduction *to* vision becomes also a reduction *of* vision: "The area of visibility in which observation is able to assume its powers is thus only what is left after these exclusions: a visibility freed from all other sensory burdens and restricted, moreover, to black and white."[7]

One might observe that this defining of observation is a kind of perceptualization of "geometric method" as then understood. This "semiotics" of vision clearly would belie what Husserl would claim about the perception of plena, but the potential role of instrumentation to transform this situation is even more strongly discounted by Foucault:

> It may perhaps be claimed that the use of the microscope compensates for these restrictions; and that though sensory experience was being restricted in the direction of its more doubtful frontiers, it was nevertheless being extended towards the new objects of a technically controlled form of observation. In fact, it was the same complex of negative conditions that limited the realm of experience and made the use of optical instruments possible.[8]

I argue against and invert the Foucaultean model later. However, one could clearly point out that it was the age-old use of instrumentation, such as Galileo's use of the telescope nearly two centuries earlier, that made the new science possible. Moreover, Galileo was quite specifically aware that this new "artificial revelation" made a new world visible and publicized this fact in his piece of scientific propaganda, *The Heavenly Messenger*. Furthermore, the actual observational or perceptual situation in the seventeenth century with respect to the microscope was a situation in which distinctive coloration was limited. The use of dyes for specimens was perfected only later.

Concerning observation through microscopes, Foucault notes:

> . . . Optical instruments were used above all as a means of discovering how the forms, arrangements, and characteristic proportions of individual adults, and of their species, could be handed on down the centuries while preserving their strictly defined identity. The microscope was called upon not to be beyond the frontiers of the fundamental domain of visibility, but to resolve one of the problems it posed.[9]

While not denying that there was a paradigm shift between this early classificatory and natural history model and the later invention of "biology," I would observe that a correlation remained between the reductionist model being employed and the technical capacity of the instruments being used. The model of comparative anatomy and function began to occur only later, when greater degrees of the microstructure were enhanced by dyeing processes which made them dramatically visible. In short, the relationship between instrumentation and Foucault's *epistemes* may have been much closer than Foucault himself could appreciate. The point I am making, however, is that at no time do the new philosophers of science make the role of instruments of scientific technologies thematic. There remains in each a certain vestigial preference for either the purely conceptual or, at most, for aspects of perception apart from its possible material embodiments through technologies.

While each of the new philosophers of science gives a minimal role to science's concrete embodiments, the technologies of instrumentation remain of secondary importance. They are cast under the umbra of a vestigial Platonism. That is why, in conrast to the philosophy of science, a more direct focus upon these material dimensions in a philosophy of technology is called for. In order to create a new gestalt, a deliberate shift toward such phenomena as instrumentation is needed. To this degree, philosophy of technology stands in contrast to its seeming near relation, the philosophy of science.

Historically, such a shift may be detected in the work of an important ancestor to each of the European thinkers already mentioned. Philosophy of technology, in its contemporary development, may be said to have roots in the work of Martin Heidegger. It began to form as early as 1927 in *Being and Time* and later took more specific shape in the period around "the Question Concerning Technology" (1950s on). Moreover, Heidegger stands as an important background figure for Husserl, Merleau-Ponty, and Foucault. The shift which occurs in Heidegger's work I shall term "materialist." Ultimately, it inverted the standard view of the science-technology relation to that of technology-science. Whatever else such inversions accomplish, at the very least they do provide radically new perspectives upon the phenomena caught in the shift. With respect to this continued discussion of the state of the art, I shall take up the central features leading to this shift in perspective.

Heidegger was a younger colleague and reader of Husserl from whom he adapted the relativistic notion of intentionality. In the ontological sense, intentionality is the relationship between all consciousness and its world or domain of objects within a field. But whereas Husserl continued to interpret this relationship as "consciousness," Heidegger took the notion in a more existential direction.

Husserl spoke of an "ego" "conscious" of the "world." Heidegger, in *Being and Time,* changes this to *Dasein*-being-in-the-World. A careful reading of what is implied demonstrates the change of significance. *Dasein,* in its ordinary German sense, means existence, and Heidegger clearly wanted to retain that sense. But in another way, he also wanted to add a technical meaning which he derived from his habits of etymologically literalizing important terms. Dasein, broken into its components, means literally "here-being" (Da-sein). This *existentialization* of experience is one which pushes what had been a kind of self-aware, knowledge-focused consciousness toward a spatio-temporal, concrete bodily sense of *position.* Here-being is the place-time I occupy, from which I experience a surrounding world or situation. Formally, in the technical language of the work, the formula is one which may be characterized in the following way:

Dasein-being in-World.

This tripartite relation contains what I have called phenomenological relativity. No element of the relation may be separated or divorced from the other; each term is, in Heidegger's language, *equiprimordial.* Furthermore, there is a correlational reversibility to the relation. If World is to be understood, it is to be from the implied *positionality,* literally the being-here of Da-sein.

Read thus, there is already conveyed a distinctly "material" sense to being human. The human being always finds himself or herself *already bodily in* a situation, *in* a World. Moreover, this existential "in" becomes the primitive for all other "ins" which could be abstracted or derived from it. The geometrical "in" as a dot within a circle is a derivative "in." Heidegger's analysis is to be the explication of the dimensions of that being-in-a-situation.

Finally, world is not only all that is contained or experienceable, but is that which surrounds one, which locates the very "here" which a human "is." All of this continues, albeit in what has been called an *existential* interpretation, the correlational relativity previously adapted by Husserl.

While this thrust is both clearly phenomenological and distinct from the older traditions, the analysis which emerged remained tainted by several of the tendencies of pre-twentieth-century philosophy. Two of these are worth noting, precisely because they point to the vestigial Platonism I have also found associated with the new philosophy of science. The first of these may be located in the habitual way philosophers of both the Modern and then of the Critical era tended to interpret the furniture of the world. The world in the transcendentalist traditions—from Descartes through Kant—was composed of objects; better, *objects of knowledge.* Granted, the particular interpretations of these objects varied: extended substance (Descartes) or constituted phenomena (Kant). But some version of an object characterized by some combination of qualities predicated of it made up the basic furniture of the world. In the contemporary era, the empirical knowledge of such objects was, of course, increasingly, science.

If one were to do a phenomenologically relative interpretation of this tradition, one would have to relate *what* is known—the totality of objects of knowledge—of *how* it is known, by some activity. The answer would be that the world of known objects implies (reflexively) a knower, and the knower of this ''World'' would, in all likelihood, be the disengaged *observer* characterized, for example, by Foucault. In the Modern era this perception is a bare and reduced perception.

The second tendency of Modern-through-Critical philosophy was to be foundationalist. Philosophy followed an architectural metaphor in which the foundation ''grounded'' that which was built upon it. The foundation was the founding stratum; that which was built upon it was what was founded. As Richard Rorty has often pointed out, both of these tendencies have been radically called into question in the late twentieth century—but in the earlier decades there were hints concerning the demise of this paradigm. Both Husserl and Heidegger—the latter only in early works—retained the architectonic of foundationalism.

Heidegger, however, as early as *Being and Time* rejected and inverted the epistemological thrust of Modern philosophy. And that is precisely where the praxis-perception tradition first takes its turn toward a philosophy of technology.

Following Husserl's strategy of beginning with that which is more immediate and familiar, Heidegger proposed to examine *everydayness* for its hidden existential and ontological implications. When he does so, he concludes that explicit acts of knowledge are not what characterize most human activity. Rather, our daily action is pragmatically actional; in its ordinariness it is, from the beginning, involved with equipment (technologies):

> The kind of dealing which is closest to us is . . . not a bare perceptual cognition, but rather that kind of concern which manipulates things and puts them to use; and this has its own kind of "knowledge". . . . Such entities are not thereby objects for knowing the "world" theoretically they are simply what gets used, what gets produced, and so forth.[10]

But this is not just a simple observation concerning what characterizes much or even most human activity. Rather, it is the first hint of the shift of perspective by which Heidegger inverts the long primacy of objects of knowledge as the primary constituents of the world of transcendental philosophy. In this *tool analysis* he argues that not only are such praxes closer to us, but that only by a kind of rupture in this familiar interaction with the environment does something like an "object of knowledge" arise.

Heidegger argues that there are several ways in which humans relate to their environment, and at least two of these may be contrasted. One such relation is the way humans engage the environment via the *ready-to-hand.* This is the pragmatic relation which implicates equipment or tools (technologies). It is an *engaged,* but also ordinary, relation the the environment. The second relation is—in a limited and special sense—"disengaged," and is essentially the knowledge relation, a relation to that which is *present-at-hand.* Heidegger argues that this second way of relating to things in the environment arises from and is dependent upon the priority of the first or pragmatic relation. (Here can be seen the seeds for the later Husserlian development of the lifeworld. Indeed, I would argue that the *Crisis* was a response to *Being and Time,* an example of an older master learning from a younger colleague to whom, in typical fashion, he does not give credit.)

If this is so, Heidegger must show why the traditions of Modern and Critical philosophy overlook his now inverted set of priorities. Heidegger's answer takes the form of a phenomenological archeology, an archeology which retrospectively may seem to lie behind the concepts of both the late (lifeworld) Husserl and again the epistemes of Foucault.

In its turn, Heidegger's famous tool analysis relies upon Husserl's earlier analysis of perception. To perceive something is never to perceive a bare thing, it is always to perceive a thing within and against its background. Minimally, there is a figure/ground relationship between a thing and its field. Thus, there is no "thing-in-itself"; there is always only a *situated* thing. Heidegger adopted this insight and applied it directly to the tacit activity which occurs in the everyday use of equipment:

> Taken strictly, there "is" no such thing as *an* equipment. To the Being of any equipment there always belongs a totality of equipment, in which it

> can be this equipment that it is. Equipment is essentially "something-in-order-to". . . . A totality of equipment is constituted by various ways of the "in-order-to," such as serviceability, conduciveness, usability, manipulability.[11]

Human pragmatic action, then, has the same structure as phenomenological perception. But strictly speaking, this perception is not "Cognitive"; it is, rather, actional. Just as one may not be *explicitly* aware that anything one sees is necessarily relative to a field—although, after variations upon gestalts are taken, one does become explicitly so aware. This unawareness also applies to involvements among tools. One may be only tacitly "aware" that tool-things are what they are by their belonging to some pragmatic context: "As the Being of something ready-to-hand, an involvement is itself discovered only on the basis of the prior discovery of a totality of involvements. . . . In this totality of involvements which has been discovered beforehand, there lurks an ontological relationship with the world."[12]

It is clear that Heidegger is implying something quite fundamental about the dimension of everyday, technologically implicated praxis. It is, in the earlier foundationalist scheme, the inversion of knowledge and action of the present-at-hand or epistemological object and the ready-to-hand or equipmental object.

Heidegger's often analyzed example is the hammer. A hammer is what it is—not first as an epistemological object, a substance which has such and such a weight or color or extension and only later is recognizable as a hammer—rather, it is first an embodiment which extends some human activity into its pragmatic context within an immediate environment. But not only is this preliminarily counter-intuitive, it is complicated by the way the hammer occurs within the actional context.

In use, when the hammer *is* dynamically its being-a-hammer, its cognitive properties are not only secondary, but as an object, the hammer may be said to "withdraw":

> The peculiarity of what is proximally ready-to-hand is that, it must, as it were, withdraw in order to be ready-to-hand quite authentically. That with which our everyday dealings proximally dwell is not the tools themselves. On the contrary, that with which we concern ourselves primarily is the work.[13]

When hammering, if attention is directed cognitively to or at the hammer, the result is usually the "wrong nail" accident. In use, tools have their distinctive kind of being in the dynamic sense and they cease to be primarily "known" objects. The hammer, in use, becomes "transparent."

There is a second kind of relativity to the tool: Its dynamic being is contextual—it belongs to a tool-context. With the hammer, that context includes the nails, the shingles, the carpentry project, etc. A hammer isolated, alone, or as a "thing-in-itself," would only vestigially be a hammer. Heidegger takes this essentially phenomenological insight a considerable step further. The tool-context contains, at least implicitly, a way of relating to an entire environment and with it an implicit "world."

> Any work with which one concerns oneself is ready-to-hand not only in the domestic world of the workshop but also in the *public* world. Along with the public world, *the environing Nature* is discovered and is accessible to everyone. In roads, streets, bridges, buildings, our concern discovers Nature as having some definite direction. A covered railway platform takes account of bad weather, an installation for public lighting takes account of the darkness. . . . In a clock, account is taken of some definite constellation in the world system. . . . When we make use of the clock-equipment, which is proximally and inconspicuously ready-to-hand, the environing Nature is ready-to-hand along with it.[14]

Through the ready-to-hand, the wider environment is encountered. It is relative to the human world of praxis and perception. One can easily see that this praxical perceptual dimension of human experience is manifest in some way within all human communities. It occurs both *without* science in its contemporary sense as well as *within* science. In this respect, "technology" precedes or is broader than an explicit science. While this implication of *Being and Time* was not solidified until later, one can appreciate the shift of priorities which appears here. It is also worth noting that even if tacit, the accounting of the technological context implies a certain "view" of nature. The technological context is, anticipatorily, a certain possible way of "seeing."

The Heideggerian inversion replaces the Modern and Critical eras' "observer" with a pragmatic and existential human "actor." This actional being is, moreover, a materialized or existential being. Finally, this materialized being is also peculiar, in that he or she is technologically involved with and extended into his or her immediate environment. The Heideggerian existentialization of the human being simultaneously materializes and technologizes action. It is a distinctly non-Platonic perspective.

Yet, while action may be said to be foundational in the context of the early works, it does not lack implication for knowledge. Knowledge is both implicated in, and arises from, the praxical:

> Our concernful absorption in whatever work-world lies closest to us, has a function of discovering; and it is essential to this function that, depending upon the way in which we are absorbed, those entitles within-the-world which are brought along in the work and with it . . . remain discoverable

> in varying degrees of explicitness and with a varying circumspective penetration.[15]

If the world is "discovered" through the praxical, the narrower sense of knowledge as epistemological knowledge is derivative. Here the inversion of action and knowledge is taken a step further. That which is ready-to-hand (actional) *founds* that which is present-at-hand (object-like). The derivation which Heidegger uses to establish this inverted relationship is of interest. The hammer returns:

> The kind of being which belongs to these entities is readiness-to-hand. But this characteristic is not to be understood as merely a way of taking them, as if we were taking such 'aspects' into the 'entities' which we proximally encounter. . . . To lay bare what is just present-to-hand and no more, cognition must first penetrate *beyond* what is ready-to-hand in our concern. *Readiness-to-hand is the way in which entities as they are 'in themselves' are defined ontologico-categorically.*[16]

So long as one is using the hammer within its work context, the hammer-as-object "withdraws" and the context of relations of the work project may remain both transparent and familiar but tacit. But what happens if the hammer is missing, or is broken, or ceases to function? Its phenomenological transparency is then changed, it becomes *opaque,* and the hammer may become an "object." That is to say, as a result of withdrawal, the hammer, even if missing, becomes not the means of achieving the work but an *obstacle* to its attainment, that which stands one against one. Using Kuhn's language, this would not be a theoretical, but a *praxical* anomaly. Heidegger puts it this way:

> Anything which is un-ready-to-hand in this way is disturbing to us, and enables us to see the *obstinacy* of that with which we must concern ourselves in the first instance before we do anything else. With this obstinacy, the presence-at-hand of the ready-to-hand makes itself known in a new way as the Being of that which lies before us and calls for our attending to it.[17]

In short, through the praxical anomaly the explicit occasion for the "object" to emerge may occur:

> *When an assignment has been disturbed*—when something is unusable for some purpose—then the assignment becomes explicit. . . . When an assignment to some particular "towards this" has been circumspectly aroused, we catch sight of the "towards this" itself, and along with it everything connected with the work—the whole workshop—as that wherein concern always dwells.[18]

This derivation of the occasion of "knowledge" makes the totality of the objects of knowledge not only derivative but special cases of human concern and activity. "Observer" consciousness is a particular development of actional, prior concerns. Thus, underneath the presumed disinterestedness of observation lies the engagement of praxis. But this is not yet sufficient to carry the implication of Heidegger's version of the technology-science relation to its ultimate conclusion.

However, that conclusion was drawn clearly some twenty-five years later in his essay, "The Question Concerning Technology." *Being and Time,* remained, at best, an anticipation of this possible philosophy of technology. Its possibility was founded upon the inversion of the usual understanding of the epistemological tradition. But more, the analysis of tools suggested certain paths toward the analysis of human-technology relations, which in turn open the way to a phenomenologically explicit philosophy of technology.

The simplest and most abbreviated way to show how Heidegger solidified his new philosophy of technology in "The Question Concerning Technology" is to employ a simple set of substitutions. The realm of the praxical—ready-to-hand—is thus the founding stratum of human-world relations in *Being and Time,* and entails a *technological* relation to the environment. The present-at-hand, which falls within science as a mode of knowledge, is founded upon the praxical relation.

What Heidegger does in this essay is invert the usual understanding of the relationship between sciences and technology. This is to say, if the dominant view claims that technology is applied science, then in Heidegger's version of the relationship science may be said to be a peculiar kind of "applied" technology. At the least, this is an inversion of Platonism and may in a curious sense even continue to be called an *existential materialism.* Such an inversion specifically understood as a deliberate gestalt shift may first seem counter-intuitive. Yet if phenomenology is correct that *intuitions are constituted,* not simply *given,* then there must be a way to reverse the standard intuition.

First, the Heideggerian inversion: Science, rather than being the origin of technology or technology as the application of science, becomes the *tool* of technology.

> It is said that modern technology is something comparably different from all earlier technologies because it is based upon modern physics as an exact science. Meanwhile we have come to understand more clearly that the reverse holds true as well: modern physics, as experimental, is dependent upon technical apparatus and upon progress in the building of apparatus.[19]

This is to say that science is necessarily embodied in instrumental technologies. This is a stronger claim than that made by any of the new philosophers of science. But that is only a preliminary point; ultimately, Heidegger claims that physics is dependent upon technology in a much more basic way. *But this may be seen only if both science and technology are discerned to be ways of seeing.*

> Modern science's way of representing pursues and entraps nature as a calculable coherence of forces. Modern physics is not experimental physics because it applies apparatus to the questioning of nature. The reverse is true. Because physics, indeed already as pure theory, sets nature up to exhibit itself as a coherence of forces calculable in advance, it orders its experiments precisely for the purpose of asking whether and how nature reports itself when set up this way.[20]

Under and behind the perspective of modern physical science lies a deeper relation to nature or the surrounding world. This is precisely the dimension of existence previously noted in *Being and Time,* the praxical relation to nature as the source of both what is to be modified by work and what is to be taken account of in terms of human action and concern. This praxis, which was already made apparent within everyday life, is now transformed into *a particular technological way of seeing.*

> Because the essence of modern technology lies in enframing, modern technology must employ exact physical science. Through its doing the deceptive illusion arises that modern technology is applied physical science. This illusion can maintain itself only so long as neither the essential origin of modern science nor indeed the essence of modern technology is found out through questioning.[21]

For Heidegger, science as a way of seeing is located within and dependent upon the priority of technology as a material, existential, and *cultural* way of seeing. Or as Heidegger puts it, technology is a *way of revealing:* "What has the essence of technology to do with revealing? The answer: everything. For every bringing forth is grounded in revealing. Technology is therefore no mere means. Technology is a way of revealing."[22]

In Western culture and history, the trajectory which leads to the interaction of contemporary science and technology relies upon a particular praxical form, a particular perspective of technology. "The revealing that rules in modern technology is a challenging, which puts to nature the unreasonable demand that it supply energy which can be extracted and stored as such."[23]

Nature is seen as the inexhaustible source of "resources" which must be obedient to man's demands, or what Heidegger calls "Standing Reserve" (Bestand). For purposes here, what is most important to grasp is that Heidegger makes the human "technological" relation to the world a mode of revealing, or a way of seeing. Technology, in the deepest Heideggerian sense, is simultaneously material-existential and *cultural.*

Western or modern technology is then *the* dominant variable in an arrangement of the praxical. It is a way of seeing embodied in a particular form.

The Heideggerian perspective upon technology, however, has been both misunderstood and ignored. Bunge's claim that Heidegger belongs to the field of antitechnologists who reify technology is somewhat misguided. Heidegger no more reifies technology than Kuhn reifies science, and for the same reasons. If technology is a historically-culturally grounded way of seeing, as science in one of its manifestations is the particular paradigmatic way of a particular community's way of seeing, one can say that both have similar structures. Kuhnian paradigms are clearly not simply individual, even if some individual first proposes a new gestalt. Nor are Heidegger's more grandiose "epochs of being" individual. Both have cultural dimensions which, as culture, situate us and are not something any one can "control" as such. Culture displays certain recalcitrant features which are nevertheless clearly recognizable as human products. And that is precisely what Heidegger does with technology. Western, or as he calls it, Modern Technology, is a particular and historical, but also existential variant upon the human relation with the surrounding world.

Heidegger clearly had misgivings about modern or scientific technology. His frequently documented preference for handwork and traditional technologies over high-tech and complex technologies reveals a strongly romantic trait running through his work. But while such a prejudice may indeed damage the Heideggerian stance—as I believe it does—this does not mitigate the early insights involving technology as a way of seeing.[24]

The very characterization of this way of seeing as placing an unreasonable demand upon nature indicates this. But Heidegger, to that degree, could clearly be in tune with persons concerned about conservation and the preservation of the environment. In effect, Heidegger was arguing that the way of seeing which is implied in contemporary technology is Baconian. While this view may have been latent from the beginning, it comes to dominance only in recent times. Heidegger, mistakenly in my view, held to a strong distinction between modern or scientific-industrial technology and traditional technologies. His romantic tendencies create a certain blindness for his insight:

> Chronologically speaking, modern physical science begins in the seventeenth century. In contrast, machine-power technology develops only in the second half of the eighteenth century. But modern technology, which for chronological reckoning is the latter, is, from the point of view of the essence holding sway in it, historically [ontologically] earlier.[25]

Heidegger also held a somewhat nostalgic view which favored older, and presumably simpler, technologies:

> The revealing that rules in modern technology is a challenging, which puts to nature the unreasonable demand that it supply energy which can be extracted and stored as such. But does this hold true for the old windmill as well? No. Its sails do indeed turn in the wind: they are left entirely to the wind's blowing. But the windmill does not unlock energy from the air currents to store it.[26]

Were this an argument for the use of renewable energy sources, it might have more justifiable sense: But in a chronological or historical sense, the point is simply romantic. For if the windmill does not store and control energy, the equally ancient and simple technology of the waterwheel does.

Although perhaps not systematized, the view that the earth is some vast, unlimited source of energy and material is also very old in the Western tradition. Mining, a very ancient technique with technologies, implicitly views the earth in the same way and with environmental results which were apparent in ancient times. The current environmental crises had plenty of ancient antecedents.

Heidegger's romanticism is well evidenced in his "The Origin of the Work of Art." He chooses as an object a Greek Temple, which, like the piece of equipment, *gathers* to itself a context of involvements:

> Standing there, the building rests on the rocky ground. This resting of the work draws up out of the rock the mystery of that rock's clumsy yet spontaneous support. Thanking there, the building withstands the storm raging above it and so first makes the storm itself manifest in its violence. . . . The temple's firm towering makes visible the invisible space of air. The steadfastness of the work contrasts with the weaving of the flowing sea, and its own repose brings out the latter's turmoil.[27]

Although he analyzes a work of art rather than a technology, the same basic phenomenological structure applies. The temple, in its stability, contrasts with its surroundings and as a focus yields a gestalt which is shaped in a particular way. It does what the hammer does in revealing a world. But virtually all Heidegger's "good" examples are of such artworks, or of farming, or peasant items (Van Gogh's shoes, the workshop tools, a windmill, etc.), while his "bad" examples are of high technology (nuclear plant, steel bridges, dams on the Rhine).

Donald Hughes, with unintentional respect to Heidegger, makes the following observation:

> Those who look at the Parthenon, that incomparable symbol of the achievements of an ancient civilization, often do not see its wider setting.

> Behind the Acropolis, the bare dry mountains of Attica show their rocky bones against the blue Mediterranean sky, and the ruin of the finest temple built by the ancient Greeks is surrounded by the far vaster ruins of an environment which they desolated at the same time.[28]

Hughes, too, sees the Parthenon in a set of involvements, a context. His view is hardly romantic without necessarily denying the underlying phenomenological point about a focus of vision.

My point is this: What Heidegger discerns as the emergence of technology as a mode of revealing is not simply postscientific. Its roots lie deep within our (and others) histories. Yet even more recently and with respect to our own historical trajectory, another essentially technological historical development precedes the rise of modern science. This development has been well documented by Lynn White, Jr., the preeminent historian of Medieval technology.

White's work was directed primarily at a reinterpretation of Medieval life with respect to technology. His classic *Medieval Technology and Social Change* appeared the same year as Kuhn's *Structure* (1962). In this book, White argued that specific technological developments implicated social change (stirrup and warfare, horse-drawn plow and agriculture, etc.). But more importantly, the overall aim of his work was to show that what amounted to a virtual technological revolution was well under way prior to the Renaissance and the rise of Modern Science.

By looking at the burgeoning technology of the Medieval Period, White paints a historical picture of a rapidly changing Europe, avidly searching for inventions and hungry for power. This is particularly evident with the newly invented mechanical devices for extracting power from water and wind. By the year 983, water power was being used for fulling mills; but within a century the Domesday census revealed that there were already 5624 watermills in operation in England (a harbinger of the Industrial Revolution centuries later).[29] The windmill is referred to as early as 1180 and is common in much of Europe by 1240. The search for power in the Middle Ages utilized every source. Inventions from foreign lands were rapidly experimented with in new ways, often hardly practical but rarely overlooked. This medieval search for power laid the groundwork for later industrial technology, but it was also intricately tied to a search for knowledge. For example, in 1420 Giovanni da Fontana designed the forerunners of our robot measurers in the form of swimming fish, flying birds, and running rabbits, all linked to a plan to measure surfaces and distances in water, the air, and out-of-the-way places.[30]

During this period, one dramatic technological development which transformed the human perception of time, was the clock. In White's words, "Suddenly, towards the middle of the fourteenth century . . . [it]

seized the imagination of our ancestors. . . . No European community felt able to hold up its head unless in its midst the planets wheeled in cycles and epicycles, while angels trumpeted, cocks crew, and apostles, kings, and prophets marched and countermarched at the booking of the hours."[31] Time and the movement of the spheres were tied to a mechanical device. Thus by 1382 the universe itself began to be conceived of according to a mechanical metaphor.

> It is in the works of the great ecclesiastic and mathematician Nicholas Oresmes, who died in 1382 as Bishop of Lisieux, that we first find the metaphor of the universe as a vast mechanical clock created and set running by God so that "all the wheels move as harmoniously as possible." It was a notation with a future: eventually the metaphor became a metaphysics.[32]

His more recent works have taken account of the unique intellectual climate which encourages technological development in Europe. By the time he publishes "Cultural Climates and Technological Advance in the Middle Ages," White can claim, "The technological creativity of medieval Europe is one of the resonant facts of history."[33] What he finds is that medieval Europe was highly receptive to the use and development of technology and that several factors encouraged this. The organization and climate for order, stemming from the earlier monastic reforms, readily adapted technology: The clock, used first to establish the order of time; agricultural techniques; and labor-saving machines were all affirmatively valued. Indeed, his survey of the literature of the time finds few who fail to praise technology. On the contrary, praise of invention, machines, and their use is the rule.

Prior to our Bishop Oresmes who declares the heavens to be clockwork, one commonly finds praise and prediction concerning a glorious technological future. "Roger Bacon, 1260, pondering transportation, confidently prophesied an age of automobiles, submarines, and airplanes."[34]

This fascination and obsession with the technological stands in stark contrast to other areas of Christian civilization. Whereas the Latin West, from the monasteries on, accepted technology into the precincts of the holy—every cathedral must have a clock—the Eastern regions forbade such inventions in sacred space. Clocks must remain outside the realm of eternity, thus outside the church in the Orthodox lands.[35]

The positive evaluation of inventiveness, linked to a desire for machine power, was also accompanied by the willingness to adapt ideas and artifacts from any culture. What became the bow for our string instruments came from Southeast Asia, a Tibetan prayer wheel may have inspired the windmill, and so the list goes. In short, the Medieval Period was suffused with interest in and desire for the development of technologies.

White points out that by the late Middle Ages, at the dawn of the time of the rise of Modern Science:

> About 1450 European intellectuals began to become aware of technological progress not as a project [. . . this came in the late thirteenth century] but as an historic and happy fact, when Giovanni Tortelli, a humanist at the papal court, composed an essay listing, and rejoicing over, new inventions unknown to the ancients. . . . It was axiomatic that man was serving God by serving himself in the technological mastery of nature. Because medieval men believed this, they devoted themselves in great numbers and with enthusiasm to the process of invention.[36]

In short, what White establishes is that by 1500, a period whose image is consolidated by the technological genius of da Vinci, there is a self-awareness of technology, the process of invention, and the desire to master Nature through human artifacts.

By the year 1500, Europe already had developed some of the instrumentation so fundamental to the very investigative possibility of science in the modern experimental sense. Lenses were invented by 1050, compound lenses by 1270, spectacles by 1285, and by 1600 (Galileo's period), the microscope and the telescope were in use. Clocks, essential to measurement, began to be developed in the ninth and tenth centuries and by 1502 were widespread from cathedral to town hall to individual use.

On the industrial side, one can note that Europe is by this time covered with wind and water mills; the lowlands were being drained by wind-power, there were railways in mines, and the massive, sophisticated architecture of cathedrals, suspension bridges, and other large projects were part of daily life. Yet in spite of the now reflective obviousness of this pervasive technological achievement of the Middle Ages, White is still probably right in claiming that "the scholarly discovery of the significance of technological advance in medieval life is so recent that it has not yet been assimilated to our normal image of the period."[37]

Such an interpretation belies the view accepted by Heidegger that modern science precedes modern technology. Not only was the technology of Medieval Europe widespread, but it was sophisticatedly "machine-like" in construction. Systems of wheels, gears, and pulleys and the complexity of machine works needed only the more autonomous power source of the steam engine in order to begin the later Industrial Revolution. Here the motto that the steam engine had more to do with science than science with the steam engine takes an earlier form. Optics and the clock also may have had more to do with the rise of science than science with the rise of either of these technologies. And in that sense, technology—in Heidegger's term—not only ontologically but also *historically* precedes what we take as science.

Heidegger's apparent lack of awareness of the history of technology causes him to overlook the quite dramatic sources which are also part of the way of seeing which belongs to so much Western history. White, quite correctly, I believe, has traced this much more macroscopic perspective to the Latin West and its dominant religious sources concerning nature. Today's close linkage between science and technology could, moreover, be located at the rise of science itself. No one is surprised by the notion that knowledge is power, a view which finds both its scientific and technological connections in the work of Francis Bacon.

The very first aphorisms in his *The Interpretation of Nature* are astonishingly modern and, in effect, unite both science and technology in ways which belie the distinctions made later by both philosophy of science and philosophy of technology:

> Neither the naked hand nor the understanding left to itself can effect much. It is by instruments and helps that the work is done, which are as much wanted for the understanding as for the hand. And as the instruments of the hand either give motion or guide it, so the instruments of the mind supply either suggestions for the understanding or cautions.[38]

Condensed here is the recognition that science gains both its knowledge and its power from instruments. But more, Bacon is already arguing that science in some sense is an "instrument of the mind." Knowledge-power is a formula which is both descriptive and uniting of the science-technology relation: "Human knowledge and human power meet in one; for where the cause is not known the effect cannot be produced. Nature to be commanded must be obeyed; and that which in contemplation is as the cause is in operation as the rule."[39]

What Bacon wanted was precisely the technological science described by Heidegger as the current state of technology as a way of seeing: "Moreover the works already known are due to chance and experiment rather than to sciences; for the sciences we now possess are merely systems for the nice ordering and setting forth of things already invented; not methods of invention or directions for new works."[40]

The themes of the control of nature through knowledge, bound to the development of instruments (technologies), were thus well recognized and sought after at the outset of the scientific era. What for Bacon was a hope and vision becomes, with Heidegger, the operational reality of contemporary praxis.

Heidegger provides us then with the thought-provoking inversion of the relationship between science and technology. Science becomes, in the Heideggerian model, the necessary tool of a technological way of relating to the world. In essence, this is what Baconian science also proposed; it was as interested in changing the world as in knowing it.

But Bacon, too, has often been overlooked. Both the Baconian vision and the Heideggerian (as corrected by White) interpretation of the relationships between science and technology open the way to an interface between science and technology.

Such an interface is actually complex and deep. I contend that contemporary science—in contrast to its ancient forms—is both technologically *embodied* in its necessary instrumentation and also *institutionally embedded* in the social structures of a technological society. Here, however, I shall look primarily at the interface between science and technology in instrumentation.[41]

Part Two

The Instrument as Interface

IV. The Embodiment of Science in Technology

Two acts in the emergence of the philosophy of technology and its interface with the philosophy of science have now been played out. Chronologically, of course, the two acts could be differently construed. The "revolutions" in the philosophy of science begun simultaneously but independently by Kuhn and Foucault, occurred in 1961 and 1962. The second act, a flashback to the thirties through the fifties via the Europeans I have dealt with, chronologically preceded these later "revolutions." But we are now ready to resume a more chronological and contemporary framework as we again return to the North American scene. Here the focus will be upon the interface of the philosophy of science and of technology by way of looking at the *embodiment of science in technology.* Most explicitly, this is in the form of its instrumentation. And this, too, is a theme which has until very recently been largely neglected by philosophers.

Alfred North Whitehead did appreciate the role of instruments, even though he did not explicitly pursue the insight he himself indicated:

> The reason we are on a higher imaginative level is not because we have a finer imagination, but because we have better instruments. In science, the most important thing that has happened in the last forty years is the advance in instrumental design . . . a fresh instrument serves the same purpose as foreign travel; it shows things in unusual combinations. The gain is more than a mere addition; it is a transformation.[1]

Nor did traditional philosophy of science pay much attention to instrumentation. Its version of science—now seen from the perspectives of the new [post-Kuhnian] philosophies of science—remained both disembodied and de-institutionalized. I have tried here to point up the implicit trajectories toward a theory of embodied science, particularly in its perceptual and praxical sense, by the Euro-American and phenomenological traditions. While I have not stressed the institutionalization of science within those traditions which began to question earlier Positivist and analytic versions of philosophy of science, social praxis in its institutional forms was at least recognized by neo-Marxians and, later, by the Popperians. Here, however, the emphasis on the interface between philosophy of science and philosophy of technology remains the praxis and perception which finds its focus in the embodiment of science in instrumentation.

There is a sense in which those from the phenomenological traditions might well be dubbed "body-philosophers," for the entire epistemology of the phenomenological critique of traditional Modern epistemology lies in the tradition's neglect of the role of body. Given this perspective, it was to be expected that the concrete roles of those material entities—instruments—would play a different role in the thought of the Euro-American thinkers than in that of traditional philosophy of science, which might be called a "mind-philosophy," where "mind" is thought of in disembodied form. Thus, in this chapter I shall briefly sketch some of the directions first taken by the "body-philosophers" coming from phenomenological traditions; I shall then turn to the growing recognition of the role instrumentation has taken in revisionist Anglo-American philosophy of science. Both movements, of course, point to the selected interface between philosophy of science and philosophy of technology previously noted.

THE "BODY-PHILOSOPHERS"

What should be noted at the outset is that the North American "body-philosophers," as I shall call them, did what their European predecessors did not do—they entered technologies into the equation between the philosophy of science and the philosophy of technology. And they did this positively. Technologies, in the form of instrumentation, were to become the way in which the gap between the lifeworld and the world of science is to be eliminated. If science, concretely, is not a matter of pure theory or simply of reasoned inferences, its more positive side relating to technology is its *embodiment* in instrumentation. But this characterization of science has seldom been considered by either traditional or new philosophers of science. What was needed was an insight into technology as a necessary, rather than an accidental, embodiment for the formation of contemporary science. This insight was to be provided, *negatively,* by an American phenomenologist, Hubert L. Dreyfus. It arose in the context of a scathing critique of the entire project of artificial intelligence, a critique which was later published as the controversial and well-known book, *What Computers Can't Do* (1972).

Background work for this book actually began at the same time as Kuhn's work in the early to mid sixties. Dreyfus was commissioned by the Rand Corporation to do a study of modeling processes in artificial intelligence, a project which culminated in 1967 in a monograph, *Alchemy and Artificial Intelligence.*

The context of this critique was not yet what one could call a development of the philosophy of technology *per se,* but it did outline certain horizons of the interface between science and technology in its contemporary sense. It revolved around the issue of *instrumental embodiment.*

Although the bulk of *What Computers Can't Do* consists of a long critique of the recalcitrant Platonism and Cartesianism within both artificial intelligence and cognitive simulation circles—a critique which sharply and clearly follows the similar programs of the classical phenomenologists—it is in the emergence of what I call "phenomenological materialism" that Dreyfus opened the way to the notion of instrumental embodiment.

The argument against the current modelling processes of artificial intelligence is cast in negative form. Dreyfus asserts that in order to think, one must have (be) a body. The rationale for this assertion comes from existential phenomenology, particularly that of Merleau-Ponty. Since computers do not have (human) bodies, they thus cannot think (humanly). It is this identification of body as a necessary condition of thought which is of primary interest here.

> Adherents of the psychological and epistemological assumptions that human behavior must be formalizable in terms of a heuristic program for a digital computer are forced to develop a theory of intelligent behavior which makes no appeal to the fact that man has a body, since at this stage at least the computer clearly hasn't one. In thinking that the body can be dispensed with, these thinkers again follow the tradition, which from Plato to Descartes has thought of the body as getting in the way of intelligence and reason, rather than being in any way indispensable for it. If the body turns out to be indispensable for intelligent behavior, then we shall have to ask whether the body can be simulated on a heuristically programmed digital computer.[7]

Then turning to Merleau-Ponty, Dreyfus adapts the perceptualist emphasis to this new twist concerning the necessity of embodiment for all thinking. Noting that artificial intelligence programs are most successful in those areas of thought which are the most formalized and closed—and thus seemingly the farthest from bodily activity—Dreyfus notes that the least successful modelling programs are found in those areas which are most explicitly closest to motile behavior:

> Indeed, it is just the bodily side of intelligent behavior which has caused the most trouble for artificial intelligence. . . . what if the work of the central nervous system depends upon the locomotive system, or to put it

> phenomenologically, what if the "higher" . . . determinate, logical and detached forms of intelligence are necessarily derived from and guided by global and involved "lower" forms?[3]

It becomes clear that this is, in fact, the position Dreyfus holds. "It turns out that it is the sort of intelligence which we share with animals, such as pattern recognition . . . that has resisted machine simulation."[4] This trajectory, of course, is in keeping with Merleau-Ponty's location of intentionality itself within motor activity, within the (human) body.

Dreyfus sees evidence for this patterned, motile intentionality in the phenomenon of pattern recognition in its phenomenological form. This form has been pointed up previously as having a certain "gestalt" set of characteristics. First, its structures include an overall sense of the global. "Pattern recognition requires a certain sort of indeterminate, global anticipation. This set or anticipation is characteristic of our body as a 'machine' of nerves and muscles whose function can be studied by the anatomist, and also of body as experienced by us, as our power to move and manipulate objects in the world."[5] Second, the structure takes the shape of a figure-ground pattern: "There is the basic figure-ground phenomenon, necessary for there to be any perception at all: whatever is prominent in our experience and engages our attention appears on a background which remains more or less indeterminate."[6] Finally, within the overall global and indeterminate anticipation fulfilled perceptually in a figure-ground phenomenon, particulars take the shape they do in terms of interconnected wholes. Here Dreyfus draws from Husserl the notions of both inner and outer horizons: "Our sense of the overall context may organize and direct our perception of the details when we understand a sentence. For a computer, which must take up every bit of information explicitly or not at all, there could be no outer horizon."[7] Outer horizons are contexts which allow objects to be selectively identified. Inner horizons function similarly as wholes "larger" than sums of their parts but which appear as such within perception itself:

> To understand this, we must consider a second kind of perceptual indeterminacy investigated by Husserl and Gestalt psychologists: what Husserl calls the inner horizon. The something-more-than-the-figure is, in this case, not as indeterminate as the outer horizon. When we perceive an object we are aware that it has more aspects than we are at the moment considering.[8]

This top-down or whole-part strategy takes as its primitive that which is first the indeterminately perceived whole. "The crucial feature of this gestalt interpretation [is] that each *part gets its meaning only in terms of the whole.*"[9]

Such perceptual pattern recognition, however, is essentially related to embodiment. This is recognized by Merleau-Ponty and pointed up more sharply by Dreyfus:

> Merleau-Ponty tries to correct Husserl . . . and at the same time develop a general description which supports the Gestaltists. He argues that it is the body which confers meanings discovered by Husserl. After all, it is our body which captures a rhythm. We have a body-set to respond to the sound pattern. The body-set is not a rule in the mind which can be formulated or entertained apart from the actual activity of anticipating the beats.[10]

In the context of the critique of artificial intelligence, the thrust of this argument is negatively placed against the modelling of a "hardware"-type body which the computer "embodies." Unlike the computer, which is limited to a digital set of possibilities determined by heuristic sets of formal rules, the human body, in its perceptual and motile functions, escapes such limiting rule-behavior and operates *informally and openly:*

> The body can constantly modify its expectations in terms of a more flexible criterion; as embodied, we need not check for specific characteristics, but simply for whether, on the basis of our expectations, we are coping with the object. Coping need not be defined by a specific set of traits but rather by an ongoing mastery which Merleau-Ponty calls *maximum grasp.* What counts as maximum grasp varies with the goal of the agent and the resources of the situation. Thus it cannot be expressed in situation-free, purpose-free terms.[11]

Again, in the context of comparing computer "behavior" to human behavior, what becomes crucial for Dreyfus is comparing the human's body to that of the computer. But in this context, the more interesting point is the recognition of the essential, necessary connection between thinking and embodiment, "since it turns out that pattern recognition is a bodily skill basic to all intelligent behavior. . . . "[12]

And what is most interesting is that it is *this bodily capacity* which enables the human to escape the closed determination of artificial intelligence:

> The body contributes three functions not present, and not as yet conceived in digital computer programs: (1) the inner horizon, that is, the partially indeterminate, predelineated anticipation of partially indeterminate data . . . (2) the global character of this anticipation which determines the meaning of the details it assimilates and is determined by them; (3) the transferability of this anticipation from one sense modality and one organ of action to another. All these are included in the general human ability to acquire bodily skills. Thanks to this fundamental ability, an embodied agent can dwell in the world in such a way as to avoid the infinite task of formalizing everything.[13]

Directed against the possibilities of a machine intelligence, Dreyfus retained an ambiguity. "On the one hand, since computers clearly don't have 'bodies' . . . as man has a body . . . at this stage, at least, the computer clearly hasn't one,"[14] and, on the other, the possibility which Dreyfus partly allows, that eventually there might be some form of "wetware" or "meat machine" type of body which could be given to computers, it remains within his account that human intelligence must remain distinctly different from artificial intelligence by virtue of *bodily* difference.

Dreyfus's role in the artificial intelligence debate has been crucial, even to this date. This is the case in his most outrageous and negative characterizations of the possibilities of artificial intelligence programs. At conferences, I have personally witnessed the way in which his criticisms have sent the prime proponents of artificial intelligence programs back to their labs to try precisely whatever Dreyfus would claim could not be done. And, at least in chess playing—a closed system game, albeit characterized by very large sets of information possibilities—his detractors have gotten the better of him. However, he remains well ahead of the game in areas such as translation programs which, in spite of terminological claims to the contrary, remain limited to sophisticated mechanical dictionary approaches.

With respect to the role of embodiment, Dreyfus played his role on a double front. Embodiment, within its human context, was taken as a positive factor. Unlike the negative evaluations of embodiment in the Platonistic and Cartesian traditions, within the "body-philosophy" of phenomenology, body is seen to play a crucial role in all epistemology. On the other front, Dreyfus, although by a negative means, points to the need to consider the different kinds of embodiment which are entailed in technologies—in his case, those of artificial intelligence machines.

However, Dreyfus's role, cast in terms of the critique of artificial intelligence, could be lost insofar as the positive insight concerning the essentiality of embodiment is concerned. However correct he may be about the differences between artificial and human intelligence, no one can gainsay the power which the computer has demonstrated with respect to certain aspects of increased knowledge. Dreyfus's argument that the computer has no (human) body and that to think humanly it must imitate our "wetware" misses two important points: First, precisely because the computer's "body" is different from ours, its "thinking" must also differ from ours. What may be missed in the debate between artificial intelligence proponents and Dreyfus is that *precisely this difference which may be of most interest.*

Historically, science itself has frequently fallen prey to its own metaphors. Terms such as "meat machines" and "wetware" are currently used within AI discourse, but only AI fundamentalists could possibly take such metaphors seriously. Rather, as a new machine the computer offers different options for the instrumental embodiment of science. In the same historical sense, arguments from analogy not only fail, as Hume showed, but they tend to emphasize a one-sidedness on the part of similarity rather than dissimilarity. Yet, once freed from the matrix of analogy, dissimilarity often proves to be the more suggestive for knowledge-gathering trajectories.

Secondly, the emphasis which emerged out of the entire AI-Dreyfus debate retained a concept of the computer as some kind of independent entity. Technologies, however, are what they are *in use and in relation to users.* This was a primary point emphasized by Heidegger in his claim that there is no such thing as "a" technology or tool apart from its context of involvements and referentialities. Dreyfus himself was well aware of this and applied the same concept to the way we perceptually and linguistically understand a world. He failed, however, to apply this same insight to human-computer use and thus failed to see a second type of possibility for the phenomenological analysis of AI, the possibility of seeing the computer as *relationally linked to a human-technology context.*

Traditionally, the history of science has only indirectly recognized this crucial instrumental role. The role of mathematization as a new form of theorizing, linked to what is usually termed *experimental* science, is given to the instrument—but as a kind of "mere means." In part, this underestimate of the role of the instrument arises out of the misunderstanding that such technologies are "neutral," when in fact they are *non-neutral.*

This was a major argumentative point of *Technics and Praxis,* in which I tried to show that technologies, once taken up, modify and transform the "worlds" which are revealed through them. It is here that both the analogues and the disanalogues between machine-mediated and direct perceptual experience become crucial. Returning to optics, telescopes and microscopes do indeed extend and magnify what can be seen, but they also reduce it by transforming the field and depth within which the object domain appears. I argued that essential to all such perceptual technologies is an ambiguous magnification/reduction structure. This structure, furthermore, may be found in different forms in all technologies, thus exhibiting the transformational non-neutrality of such technologies.

Such non-neutrality, with respect to knowledge-gathering in science, is not itself intrinsically negative or positive. Both the magnification and the reduction can serve different, but related, purposes. The price paid for visual extension may be, in one sense, a reduction of depth of field. But inversely, such a reduction of the depth of field can also serve other positive purposes. It is often the case that in precisely such differences from human perception, new phenomena also are revealed.

For example, within the realm of visual representability, the contemporary use of infra-red photography from satellites is an interesting form of instrumental "phenomenological variation." Infrared photography contrasts and reveals foliage and organic matter in much greater detail than would be the case with normal coloration. But it is also disanalogous from ordinary vision. Better still are the new uses of "false color" for other newly revealed phenomena.

In false-color projects, the deliberate introduction of color makes visible that which ordinarily would be either invisible or too lightly contrasted to notice. More extreme examples of disanalogous forms of vision are even more suggestive. In today's military technologies, far and near heat sensors linked to light-enhancing nighttime telescopy reveal a visual heat-profile of such entities as tanks or humans which strict eye-analogues never would show.

These examples illustrate one line of possibility for the human-technology interface when it is regarded as a relational or symbiotic unity. It was that relationality and its forms which I stressed in my earliest entry into philosophy of technology, "A Phenomenology of Man-Machine Relations," originally appearing in 1975. I argued that technologies should not be conceived of abstractly as entities-in-themselves but relationally and symbiotically with their human users. Later, in the first four chapters of *Technics and Praxis,* I developed a phenomenology of scientific instrumentation to show how contemporary science is embodied in its technology—instrumentation.

Following the same perceptualist traditions as Dreyfus, I first took account of those human-technology relations which I call *embodiment relations.* These relations specifically extend and transform human bodily and perceptual intentionalities. Unlike the computer—which, as we shall see, falls into a different kind of human-technology relation—many scientific instruments do incorporate perceptual and bodily praxis. And these do not need to be as complex as the computer.

Simple examples are, of course, those which dominated the days of early modern science: the optical instruments of the telescope and the microscope. Such instruments extend the human senses, and in the context of early modern science, made its first distinctive embodiment as a thoroughly technologically embodied science possible. Galileo's use of the difficult and primitive telescope provided him with an "artificial revelation" of phenomena never before seen nor suspected. Along with its use of a unique form of mathematization, early modern science was to henceforth differ from all its predecessors by being *instrumentally embodied in technologies.* The resulting cosmologies were different from all previous "eyeball" cosmologies, even though they would retain perceptual representations.

One overall result of the taking up of instrumentation was the reconfigured emphasis upon both the micro- and macrodimensions of things. Magnification of the heavens (macro) and of the sub-perceptual (micro) would transform the very fields which could be investigated. Here was already a latent trajectory resultant from the taking up of instrumentation, which today frequently dominates the fields which become defined as frontier research areas.

In such embodiment relations there is, of course, no temptation to think of the machines used as "intelligent," as seems to be the case in AI contexts. They remain "extensions" of human perceptual and bodily activity because they so obviously belong to human-technology contexts. I argued, in *Technics and Praxis,* that a second group of relations does not extend or mimic sensory-bodily capacities but, rather, linguistic and interpretive capacities. These relations I called *hermeneutic relations.* Historically, the invention of writing and its technologies are the most dramatic examples of human-technology uses in the hermeneutic dimension. There does remain a metaphorical analogue with embodiment relations within hermeneutic relations. If embodiment technologies allow that which is fully perceptual to be made present in a kind of instrumentally mediated partial transparency, so too do hermeneutic relations allow for their own kind of "transparency." But that transparency is primarily linguistic-interpretive rather than directly or mediately perceptual.

In embodiment relations the technology—instrument—is taken into perceptual and bodily experience, as with Merleau-Ponty's blind man's cane or Heidegger's hammer. The technology becomes a "part" of my now extended bodily experience of the world. But in hermeneutic relations, the technology is much more "text-like" and retains a sense of quasi-otherness not present in embodiment relations. The hermeneutic "transparency" is thus like that of textual transparency. It is still referential but being "read-through."

In hermeneutic relations the technology is not so much experienced-through as experienced-with. The perceptual act directed *toward* the technology is a specialized interpretive act. Thus, whether what is being read is a text, a map, an instrument with numbers or a scale, or the dynamically appearing printout of the computer, the "transparence" is not isomorphically perceptual, although it occurs within a perceptual context.

Hermeneutic relations, insofar as they are analogues of language activities, are to human-technology relations what writing is to speech. And, as such, they are equally transformative.

It is at this juncture that one may return to the AI-Dreyfus debate in a different way. Within the very relational experience of humans with computers there are two aspects which provide a temptation to anthropomorphize the computer, as in AI projections: (a) First, in a hermeneutic relation, a technology always appears as a quasi-other in that the experience is focused upon and directed at a technology. And

(b) insofar as the computer specifically mimics linguistic-interpretive activity, it displays a certain analogocity to at least some of human activity in a way unlike other technologies. In short, there is within this particular set of hermeneutic contexts an inclination to take the computer as a *similacrum.*

What could be missed, however, is the underlying potential for computer use to stimulate quite different directions of knowledge gathering. I shall suggest two: (1) The first lies in an entirely different mode of perceptual patterning than occurs within human thought patterns or in the way in which these are generated. In recent work in mathematics—what is becoming known as "experimental mathematics"—the use of a unique adaptation of the computer to pattern generation has appeared. Numbers and number series—the "thought content" of computers—are fed into the computer to generate graphics. When this is done in relation to such mathematical phenomena as fractals or chaos phenomena enhanced by false color, patterns emerge which exhibit, on the one hand, features of artificially generated terrains strikingly similar to actual geological terrains, thereby suggesting a role for such mathematics in relation to natural phenomena, and on the other hand, repetitions of patterns in random or infinite projections which suggest a certain patterning to even "chaotic" phenomena. In this return to perception—albeit through a completely artificial means—the computer shows promise for an entirely new embodiment of what was once the most abstract and "non-perceptual" science in parallel to the instrumental embodiments of many earlier technologized sciences. It should be noted that this use of computer-generated patterns reverses the problem of the pattern recognition programs which attempt to replicate what are taken to be perceptual patterns by reducing them to mathematical progressions. Here, mathematical progressions are used to generate patterns (which then can be interpreted in the gestalt fashion of insightful human perceptions). The production of such repeated patterns as the "Mandelbrot" configuration, which turned out to repeat itself in a wide variety of chaotic projections, could not be called so much a product of "computer intelligence" as the making available in a *perceptually intelligible* way that lay hidden in the complexity of number projections. Here the very clumsiness of binary "thinking" is overcome in a different form of gestalt, suggestive to informed perception.

(2) The second line of development lies closer to AI applications but with a different twist. Pattern recognition programs which have proven most successful in application are often those in which computer "perceptions" again differ greatly from their human counterpart perceptions. This is often the case with so-called "expert" programs such as those used in medical diagnosis. What sometimes emerge as striking and of innovative significance are the correlations of

diagnosis programs with series of questions which yield results quite unexpected from the skilled sedimentations of the physicians.

What both these developments show is that in spite of the inventor and his or her intentions, the human-technology interface can yield both analogue and disanalogue results, both of which may lead to new knowledge if critically appreciated. However, this is a symbiotic relationship, not one in which the technology-per-se does the discovering. The symbiosis belongs to the combined human-technology interface and exists in a hermeneutic context.

In even more recent history, this hermeneutic role of instruments has been appreciated and developed independently by two other philosophers of science: Patrick Heelan, in *Space Perception and the Philosophy of Science* (1983); and Robert Ackermann, in his *Data, Instruments, and Theory* (1985).

Heelan's approach, like that of myself and Dreyfus, stems from his background in Euro-American philosophy and owes much to classical phenomenology. Heelan's recognition of the role of instruments in actually making perceptually present a "scientific" world goes back to his article, "Horizon, Objectivity and Reality in the Physical Sciences" (1967), and is thus contemporaneous with the work of Dreyfus. (My own work on the phenomenology of instrumentation was in part a response to Heelan. In our discussions, which began when we became colleagues at Stony Brook in 1970, there has been a long and friendly dialogue on this use of phenomenology in the interpretation of instrumentation.)

Heelan is a rarity within the philosophy of science in that he is genuinely a multidisciplinary-trained scientist and philosopher, with doctorates both in physics and in philosophy. In that context, as well as from the Euro-American traditions he inherited, he has been more sensitive to the *praxis* of actual science than most mainline philosophers. Thus, he developed an early appreciation for instrumentation's crucial role in scientific discovery.

In his *Space Perception and the Philosophy of Science* (1983), Heelan elaborates on one of the most sophisticated attempts to show the way a perceptualist and praxical theory of science entails instrumental embodiment. His chosen context is a long debate within the philosophy of science revolving around presumed differences between "theoretical" entities and perceived or "observed" entities. In the process, he not only develops more fully than the classical phenomenologists a praxis-perception model of knowledge but sees instrumentation—technology—as the means whereby this model is verified.

In Heelan's account there are three issues which relate to his version of a praxis-perception model of science: The first echoes and arises directly out of the history of the attack upon Modern epistemology through phenomenology. Heelan, like Husserl and

Merleau-Ponty before him, rejects the Modern distinction between primary and secondary sense qualities and substitutes the phenomenological interpretation of bodily intentionality as the relation between "subjects" and "objects." In this, he remains phenomenologically orthodox.

This in turn means that Heelan must—and does—reject the form of transcendentalism which adheres to the dominant view in the philosophy of science regarding both objectivity and the implied notion of a disembodied, ideal observer. By contrast, all observation must be interpreted relationally as observation taken by embodied—and hence, positioned and finite—observers. Furthermore, in this view science becomes part of the overall cultural lifeworld, as it did in Husserl and Merleau-Ponty. This places Heelan among the new philosophers of science noted in the narrative of the previous history.

Where Heelan begins to develop his own insights, however, is in relation to his desire to retain a form of *scientific realism,* albeit of a new philosophy-of-science type. It is what he calls "*horizonal realism,*" which belongs within the aims of phenomenology, particularly in its hermeneutic form. Here he sets himself apart from both classical phenomenology and the dominant version of the philosophy of science.

At the heart of this second issue lies the confusion over the presumed *imperceptibility of scientific (theoretical) entities.* Of course, in the tradition of Modern philosophy, which revived the Democritean notion of sense and reason, entities such as atoms were supposed to be *in principle* imperceptible. But as we have already seen, phenomenology rejects this metaphysical distinction. Heelan argues that in the actual practice of science such entities are not only not in principle imperceptible but are regularly perceived *through the mediation of instruments,* i.e., technologies:

> I shall contend that the theoretical entities of science, "notoriously imperceptible and unimaginable" to the unaided senses, can nevertheless, through the use of readable technologies, become fully perceptible and as a consequence can take a place in the World of manifest objects, membership in which, in my view, gives the status of reality.[15]

The itinerary Heelan then takes becomes an attack upon the very distinction between what is observable and what is theoretical. "Within such a perspective, a distinction between *observational* and the *theoretical* probably no longer makes sense."[16] What has been previously regarded as imperceptible (theoretical) becomes in the new context a special but restricted step within scientific discovery. "The term 'theoretical,' however, can be given another sense. It can be taken to refer to one of the components of scientific activity, namely, logical and mathematical deduction, in contrast with the making of observations. . . ."[17]

In short, by means of scientific technology Heelan moves what was formerly considered to be imperceptible into the range of human perceivability. Heelan thus rejects any deep division between a "scientific world" and the "lifeworld." Those "worlds" overlap at the juncture of technological embodiment. In the process, Heelan takes the notion of the instrumental embodiment of science further than it had been taken in either classical empiricism or classical phenomenology.

The embodiment of science in instrumentation is where this step beyond the classical versions of science occurs. In Heelan's view, science, embodied in its new mediated perceptions, can bring genuinely before human perceivability precisely those "worlds" or aspects of the World which had hitherto not been perceived: "In my account, *reality* is taken in a different way: It is exactly what Worlds make manifest . . . to human perceivers; consequently, science to be realistic must have as its primary goal the exhibiting of reality structures not accessible to prescientific perception."[10] This is, of course, what the scientific instrument does, thereby expanding the previously restricted "eyeball" world into the micro- and macro-enhanced, instrumentally real, scientific world.

I shall not follow here the detailed analysis of *Space Perception* regarding the multiple meanings of perception which often confuse the issue, but shall turn in Heelan's interpretation of instrumental embodiment, which constitutes the third set of issues within his re-interpretation of the science/technology interface.

In effect, scientific perception is a highly specialized mode of perception, whether it occurs through face-to-face observations or is mediated through technologies. In a scientific observation, one can say that there is always both *perception* and *measurement.* Indeed, it is the introduction of this specialized "measuring perception" which lies at the root of much of the history of debate which Heelan enters.

Heelan's solution—which I view as something of a shortcut—is to interpret perception in a particularly hermeneutic way. Scientific instruments are both means of perception and measurement as a particular type of "readable technology." Scientific perception is thus analogous to the reading of a text—although within reading, there is a perceptual act:

> Theoretical quantities become known by attending to the response of appropriate empirical procedures, such as the use of measuring instruments. There are, however, two ways of attending to the response of an instrument: the instrumental response can be used in a deductive argument to infer a conclusion, say, the value of a quantity (provided one is experienced and skillful in the use of the instrument).[19]

We have already seen that in the first sense, one may retain a special use of "theoretical" as a pointing to or conjecture about what is not yet within the range of perceivability. But in the second sense, what becomes "observable" is that which now can be perceived—but as "read through the instrument." Heelan's version of perception is the specialized perception of a "reading."

Since measuring remains essential to the special revelation of the World in science, Heelan's version of a "measuring perception" comes close, on the one side, to being a perception essentially or necessarily instrument-embodied:

> . . . the perceiver has to be embodied in instruments or an appropriate technology. These instruments or this technology, I shall argue, are essential to the isolation of an object from its background and for the identification and recognition of any particular molecular array: they define what an array of this kind is.[20]

Insofar as some instrument as a "ruler-analog" may be used—even in eyeball observations—the measurement side of scientific observation may, indeed, be essentially technologically mediated.

But on the other side of the equation, Heelan collapses what had previously been bodily perceivability, as in Husserl and Merleau-Ponty, into the special kind of hermeneutic perceivability of "readability." Heelan argues that in the readability of skilled instrument use, there is the full characteristic set of qualities belonging to perception: "I . . . claim that this 'reading' is a perceptual process, since *it fulfills all the characteristics of perceptual knowledge.*"[21]

One example used is that of reading a thermometer:

> In "reading" a thermometer, say, one does not proceed from a statement about the position of the mercury on a scale to infer a conclusion about the temperature of the room by a deductive argument based on thermodynamics; of course, one could, but then one is not "reading" the thermometer.
>
> To the extent one "reads" the thermometer, the thermodynamic argument remains in the background, being merely the historical reason why thermometers came to be constructed in the first place. One can "read" a thermometer, however, whether or not one knows anything formal about thermodynamical theory. Provided the instrument is standardized, and so can function as a readable technology, the instrument itself can define the perceptual profiles and essence of temperature.[22]

This kind of measuring perception, Heelan claims, is "essentially both *hermeneutical* and *perceptual.*"[23]

In order for science to reveal to perceivability its world, the instrument or technology must function like perception in its direct form. The perceptual referents, through technologies, retain their realism of presence to an observer. And in Heelan's second sense of instrument use, that is precisely what a scientific instrument does:

> A scientific instrument, always a macroscopic device, can, for example, serve an investigator in two ways: (1) indirectly and inferentially, for detection purposes—this usage is theoretical—and (2) as a "window" opening directly on the hidden processes and structures of nature—this usage is observational.[24]

To this, Heelan adds a stronger claim: Measuring perception is *only and specifically that perception which is mediated through technologies:*

> Now the position I have been defending is that theoretical states and entities are or become directly perceivable . . . because the measuring process can be or become a "readable technology," a new form of embodiment for the scientific observer. In this view, the term "observation" no longer means *unaided* perception. It implies that theoretical states and entities are real and belong to . . . "the furniture of the earth," *because they are perceivable in the perceiver's new embodiment.*[25]

Ultimately, this definition must be seen as a narrow one, since it excludes forms of observation unaided by instruments. I suspect that here Heelan is being guided implicitly by his own preference for physics as the "foundational" science; if this is so, he remains to that extent part of the traditional philosophy of science. Here then is a very strong claim relating to the necessity of the technological embodiment of science in instruments as the location of a science/technology interface.

As strong as the claim is for the necessity of a technological embodied science, there remains within this position a problem relating to its inversion of reading and the perceptual act. Not all perception implies "reading" except in a highly metaphorical sense. Nor does all perception—even in science—imply the specificity of the measuring act. Perhaps only in what Heelan continues to take as the paradigm science—"theoretical physics"—with its field of constituted sub-(eyeball) perceptual entities does this fully hold.

There remain interesting areas of overlap between what could be called the public objectives of perceptual phenomena and their scientifically constituted counterpart contexts—as Heelan shows brilliantly in the example of color judgments.[26] Nor does Heelan's unnecessarily narrow account work nearly as well in those observational sciences which continue to harbor much wider fields of direct bodily or non-instrument-mediated observation—as in many of the ecological sciences. (Admittedly, measurements are taken in such sciences as well, although the revealing of entities themselves are not necessarily dependent upon instrumental mediation.) In my view there are many areas in science where a meaningful distinction between embodiments (bodily-sensory relations) which remain isomorphic with perception, and hermeneutic ones which are specialized as 'readings' but which are not isomorphic with bodily perception. Heelan's account tends to collapse the two into hermeneutic ones alone. Yet Heelan's is one of the strongest cases to date for demonstrating the essential interface between science and technology at the very heart of the knowledge-gathering enterprise.

What then emerges from this brief survey of three contemporary "body-philosophers" in the phenomenological sense is a positive and essential role for technology within science. And while the appreciation of what I shall later call "instrumental realism" arises in a direct trajectory from classical phenomenology, its placement within the philosophy of science/philosophy of technology interface is unique to the North American contemporaries.

THE "MIND-PHILOSOPHER" TRADITION

It would be a distortion of the history of philosophy of science to ignore movements similar to those above, but which come from the phenomenological traditions within the Anglo-American traditions. But the way in which the role of instrumentation arrives as a crucial factor within that tradition is different from the way it appears in the Euro-American traditions. With the latter, I have tried to show that the role of technology in science is recognized by means of a "body-epistemology" which animates perceptualist phenomenology. No such epistemological tradition exists within the positivist or analytic traditions. Still bound—with some notable exceptions—to models of neo-empiricism, much dominant philosophy of science spent its immediate post-Kuhn years launching attacks upon what they thought to be Kuhn's irrationalism and relativism.

If Kuhn's approach remains external to the dominant tradition's view, internal to the tradition was the debate about realism and antirealism. I have already shown that those from the phenomenological traditions adopt what could be called a modified or "instrumental" realist position. A similar position is taken by Ian Hacking in his *Representing and Intervening* (1983), a book published simultaneously with Heelan's *Space Perception.*

Hacking's context partially overlaps Heelan's concerns. In the first half of the book *Representing,* the central issue remains the positivist and analytic philosophy of science concern over realism/antirealism issues. But in the second half of *Intervening,* Hacking focuses on the role of *experiment,* which he considers to have been largely neglected by the dominant traditions of philosophy of science because of its almost exclusively theoretical and—in the case of both Positivist and analytic forms—linguistic or propositional emphasis. By recalling experiment as a central function in scientific discovery, Hacking necessarily implicates practice, technology, and perception (by way of observation).

Compared to the previously noted philosophers within the

phenomenological traditions, Hacking's concern with perception is, at most, indirect, but there is little "perception talk"—to use analytic jargon—and his implicit theory of the senses remains quite traditional. Nevertheless, the role of both observation—hence, "seeing"—and the instrument—hence, technology—become central. Indeed, Hacking's revival of interest in the experiment preceded a whole series of works on experiment and instrument in the dominant traditions.

Hacking is clearly revisionist with respect to the dominant traditions, even if his revisionism remains common-sensical and eclectic in approach. Throughout the polemics of *Representing and Intervening,* he takes an explicit antipositivist stance, while explicitly excluding any consideration of phenomenological traditions in the philosophy of science. Noting that there is a change in what counts as "seeing" after 1800, a change differently taken account of by both positivism and phenomenology, Hacking says: "This is the starting point for both positivism and phenomenology. Only the former concerns us here. To positivism [and, I would add, phenomenology] we owe the need to distinguish sharply between inference and seeing with the naked eye (or other unaided senses)."[27] In spite of this deliberate omission Hacking arrives at a position which shows striking parallels to that of the North American phenomenologists previously noted.

The critique which opens the way to this parallelism arises in the context of his revived interest in the experiment. By emphasizing experiment, Hacking explicitly emphasizes the praxis of science along with its theorizing. Hacking's own position vis-à-vis the theory/practice issue is eclectic. There are many possibilities for interrelations between theory and practice, all of which have occurred in the history of the development of science.

> Some profound experimental work is generated entirely by theory. Some great theories spring from pretheoretical experiment. Some theories languish for lack of mesh with the real world, while some experimental phenomena sit idle for lack of theory. Thus I make no claim that experimental work could exist independently of theory. . . . It remains the case, however, that much truly fundamental research precedes any relevant theory whatsoever.[28]

Hacking's critique of analytic philosophy of science is thus partly an issue of balance—experiment has been overlooked by too large a preoccupation with theory. "Philosophers of science constantly discuss theories and representation of reality, but say almost nothing about experiment, technology, or the use of knowledge to alter the world. This is odd, because "experimental method" used to be just another name for scientific method."[29]

However, opening the door to actual scientific practice also leads to a deeper implicit critique of a purely theory-oriented philosophy of science; it ultimately must also question the dominance of linguistic- or propositional-centered philosophy. Hacking partially recognizes this problem within the dominant views (the critique arises in a debate about observation, which necessarily implicates perception and "seeing"):

> Commonplace facts about observation have been distorted by two philosophical fashions. One is the vogue for what Quine calls semantic ascent (don't talk about things, talk about the way we talk about things). The other is the domination of experiment by theory. The former says not to think about observation, but about observation statements . . . the latter says that every observation statement is loaded with theory. . . . [30]

Hacking's own response is to return to a (common-sensical) consideration of observation itself: "If we want a comprehensive account of scientific life, we should, in exact opposite to Quine [and the fashion], drop the talk of observation sentences and speak instead about observation."[31]

This return to the context of experiment and observation as essential components of the scientific enterprise clearly entails at least indirect consideration of perception and direct consideration of praxis and instrumentation. And this is exactly what occurs within the "intervening" section of Hacking's book. I here select for exposition some of the features which precisely reciprocate those noted by our Euro-American philosophers.

One ultimate result of Hacking's stance, in the context of the realism/antirealism debate, is the adoption of a modified and modest realist position. Insofar as experimental science is differentiatable from theoretical science, it is necessarily "realist."

> Experimental work provides the strongest evidence for scientific realism. This is not because we test hypotheses about entities. It is because entities that in principle cannot be "observed" are regularly manipulated to produce a new phenomena [sic] and to investigate other aspects of nature. . . . By the time that we can say we can use the electron to manipulate other parts of nature in a systematic way, the electron has ceased to be something hypothetical, something inferred. It has ceased to be theoretical and has become experimental.[32]

But it arrives within a presented, manifest world by way of its presence within the experimental—*technologically mediated*—context.

Hacking, in a view similar to Heelan's, holds that those once-theoretical entities which come into the range of manipulatability may be considered real. But some theoretical entities will remain only inferred; and regarding these, there is room for considerable skepticism.

"Long-lived theoretical entities, which don't end up being manipulated, commonly turn out to have been wonderful mistakes."[33] Thus, both Heelan and Hacking narrow the realm of what counts as theoretical, and in contrast, give what is instrumentally observable or manipulable a larger scope within science. This, too, is the area of the science-technology interface.

However, in the process of entities coming into manipulability, the perception-praxis model of scientific discovery appears within Hacking's own view. His brilliant chapter on microscopes illustrates most vividly this need for interfacing science and technology. The puzzle is posed in typically "analytic" fashion, in a question as to whether we do indeed "see" with the microscope. Limiting his notion of "seeing" to a narrower one of direct, naked seeing, Hacking for the most part glosses over what counts as seeing in his statement of the problem, "We do not in general see through a microscope; we see with one. But do we *see* with a microscope?" And he concludes in equally "analytic" fashion with the modest, "I know of no confusion that will result from talk of seeing with a microscope."[34] Between the question about seeing with a microscope and his almost grudging conclusion that we do see lies a cautionary tale which is of interesting reciprocity with the Euro-American discussion of instrument-mediated perception.

Heelan's position regarding instrumentation is clearly an enthusiastic endorsement of a "seeing" with instruments, albeit cast in the hermeneutic terms of "reading" a "text" provided by instruments. Nevertheless, this "reading" is held to have all the qualities of a perceptual "seeing." Thus, one can say that Heelan's is a specialized but liberal interpretation of "seeing with" an instrument.

At first—and in contrast with the more restrained, linguistically bound philosophers—Hacking, too, shows a certain enthusiasm about instrument-mediated perception. "One fact about medium-size theoretical entities is so compelling an argument for medium-size realism that philosophers blush to dismiss it: Microscopes. . . . The microscopist has far more amazing tricks than the most imaginative of armchair students of the philosophy of science."[35] Hacking also recognizes that any meaningful result of instrument-mediated perception is tied to a process of learning to see: "You learn to see through a microscope by doing, not just by seeing,"[36] and that, acquired, such seeing is a matter of skill. In this sense, scientific seeing is from the beginning not merely "ordinary" seeing. "Observation is a skill. Some people are better at it than others. You can often improve this skill by training and practice."[37] These factors are stipulated as necessary for Heelan's hermeneutic perception.

Such tacit knowledge, once acquired, also is a way of seeing—doing—that needs not entail knowledge of explicit theory. "You do not need theory to use one [a microscope]. . . . Hardly any biologists know enough about optics to satisfy a physicist. Practice—I mean in general doing, no looking—creates the ability to distinguish between visible artifacts of the preparation of the instrument."[38] And Hacking recognizes some of the non-inferential immediacy of perceptual gestalts which result from instrument-mediated perception. Citing a question about whether certain dense bodies seen in an electron microscope process were entities within the bodies of the entities being investigated or "artifacts" of the instrument itself, he notes that so long as multiple processes show the same result, there is an implicit "argument from coincidence" within such observations. "My mentor, Richard Skaer, had in fact expected to prove that dense bodies are artifacts. Five minutes after examining his completed experimental micrographs, he knew he was wrong."[39]

In all these respects, Hacking is closely paralleled by the phenomenological "instrumental realists." However, there also remain distinct differences—although nuanced ones—between Hacking's stance and those of the Euro-Americans. These differences arise in part from different focuses upon the instrument. Both Heelan and I focus upon the instrument-in-skilled-use when making claims about *seeing through* an instrument; we take a fully developed instrument for granted and focus upon what is seen as a result of its development. Hacking is much more concerned with both the technical problems and developmental problems which occur within the history of the instrument, particularly with its impediments to seeing. Hacking becomes from the start more concerned with (a) how an instrument is made, particularly with respect to its theory-driven design, and (b) the physical processes entailed in the "how" or conditions of use. This means, in turn, that whatever his theory of perception, it gets cast into a causal network which allows Hacking to recognize that, while one sees "with"—as he says—a microscope, such "seeing" is also different in some sense from ordinary seeing.

The role of technology, the instrument, is an important one in science, although for Hacking it is not solely through instruments or in instrument-constituted contexts that scientific knowledge arises (as it does in Heelan's narrower, stipulative definition). But in those contexts in which much scientific observation does occur with instruments, Hacking notes that (a) "Often the experimental task, and the text of ingenuity of even greatness, is less to observe and report, than to get some bit of equipment to exhibit phenomena in a reliable way,"[40] and that (b) in the contemporary context, more and more is done through technological embodiment: "Although there is a concept of "seeing with the naked eye," scientists seldom restrict observation to that. We usually observe objects or events with instruments. The things that are "seen" in twentieth-century science can seldom be observed by the unaided human senses."[41]

Yet as Hacking recognizes, this twentieth-century habit was also part of the very beginning of Modern science, particularly in the thought of Francis Bacon:

> The word "observation" was current in English when Bacon wrote, and applied chiefly to observations of the altitude of heavenly bodies, such as the sun. Hence from the very beginning, observation was associated with the use of instruments. . . . [Hacking goes on to note some of the different uses in Bacon. Some are made with devices that aid the immediate actions of the senses.] These include not only the new microscopes and Galileo's telescope but also rods, astrolabes and the like which do not enlarge the sense of sight, but rectify and direct it. Bacon moves on to "evoking" devices that "reduce the non-sensible to the sensible; that is, make manifest things not directly perceptible by means of others which are." Bacon thus knows the difference between what is directly perceptible and those invisible events which can only be "evoked."[42]

Thus, before both positivism and phenomenology, a distinction between direct and mediated perception was recognized. In some sense, mediated perception must entail differences—the question is how and what such differences may be.

Hacking points out that the microscope became only gradually the useful instrument it is today, even for biology. He notes:

> We often regard Xavier Bichat as the founder of histology, the study of living tissues. In 1800 he would not allow a microscope in his lab. . . . He wrote that, "When people observe in conditions of obscurity each sees in his own way and according as he is affected. It is, therefore, observation of the vital properties that must guide us" rather than the blurred images provided by the best of microscopes.[43]

And it is phenomenologically true that: "Magnification is worthless if it 'magnifies' two distinct dots into one big blur. One needs to resolve the dots into two distinct images."[44] Yet phenomenologically, the very appearance of such blurs is also evidence for the as yet imperfect instrumental transparency of the technology. Were either one to simply and uncritically accept any result of instrument mediation as a transparent reference to a world or to disregard the corrective vector provided precisely by actual frustrations to instrument-mediated vision, it would be unlikely that further development would occur.

There is a second problem which faced early microscopy as well. Many of the objects in biology were small creatures with *transparent* bodies which could not be seen by lack of contrast within the living tissue. Thus unlike the telescope, which came closer to an immediate magnification of visual objects, microscopy was beset by various problems (although Bichat's objection to microscopes is not much different from that of the Aristotelians' to Galileo's primitive and difficult-to-use telescope). Thus, Hacking turns to the history and theory of microscope development to situate the way in which technology transforms ordinary vision.

With respect to the light microscope, Hacking notes, "There are about eight aberrations in bare-bones light microscopy. Two important ones are spherical and chromatic."[45] The effect of optics on the making visible of the microscopic remained serious detriments to instrument evolution until achromatic lenses could be produced. "No one tried very hard to make achromatic microscopes, because Newton had written that they are physically impossible. They were made possible by the advent of flint glass, with refractive indices different from that of ordinary glass."[46] In short, primitive microscopes retained both a kind of opacity and resolution-ambiguity such that they had not yet attained "instrumental transparency" in my sense of the word. At this stage one could not "see through" a microscope in any way analogous to naked vision. But that is not to say that development did not render "seeing through" eventually possible.

Similarly, the problem of tissue transparency was not resolved until the invention of aniline dyes for staining, as Hacking points out. "The new aniline dyes made it possible for us to see microbes and much else."[47] Here was an early use of "false color." But neither of these issues actually touches upon the phenomenology of instrument-mediated vision. They are, rather, preconditions for instrumental "transparency" in both Heelan's and my case.

This is not to say that the developmental-physical property analysis lacks interesting results; in fact, its very indirectness does arrive at several important insights differently arrived at phenomenologically. The first of these is the recognition that a *transformed, instrument-mediated* vision can, by its difference from ordinary vision, also make phenomena manifest in scientifically interesting ways.

> We do not need the "normal" physics of seeing in order to perceive structures in living material. In fact we seldom use it. Even in the standard case we synthesize diffracted rays rather than seeing the specimen by way of "normal" visual physics. The polarizing microscope reminds us that there is more to light than refraction, absorption and diffraction. We could use any property of light that interacts with a specimen in order to study the structure of the specimen. Indeed we could use any property of *any kind of wave* at all.[48]

This point is clearly consonant with my own approach in *Technics and Praxis,* where I pointed out that along with any "magnification," a technology also always "reduces" its object. The limitations of only certain properties of light in microscopy thus constitute instrumental "phenomenological profiles" upon the object or specimen in question.

Secondly and more importantly for scientific knowledge, it is precisely in the transformative "vision" of the instrument that much knowledge is, in fact, discovered. Hacking also indirectly recognizes this second point. It is by adumbrating profiles that the entity gradually gains it identity as "real." Hacking lays this at the feet of the corroboration of the entity through what he previously called the "argument from coincidence." "We are convinced because instruments using entirely different physical principles lead us to observe pretty much the same structures in the same specimen."[49] Both the adumbration of profiles and the transformational enhancement of the object are interestingly illustrated in the new *acoustic* microscope (ultrasound microscopes). By using high-frequency sound waves—and instrumentally converting these into visual displays—the acoustic microscope reveals phenomena known more to hearing than to vision. Hacking notes:

> As always a new kind of microscope is interesting because of the new aspects of a specimen that it may reveal. Changes in the refractive index are vastly greater for sound than for light. Moreover sound is transmitted through objects that are completely opaque. . . . The acoustic microscope is sensitive to density, viscosity and flexibility of living matter.[50]

And although the instrument follows the usual visualist prejudice of science in its hermeneutic "translation" of sound into visible display, the properties revealed are those which are also ordinarily revealed to human "naked" hearing.

Thus, if there is an issue between Hacking and the phenomenologists, it lies more in peripheral than in final observations. Hacking is fussy about too liberal a notion of "seeing," whereas Heelan is obviously very liberal. Yet both can agree that the kind of experimental result is a realist one, that it is arrived at by a process other than inferential thinking and is mediated by instruments, the interface between science and technology.

Hacking's cautionary note about "seeing with" rather than "through" a microscope indirectly points up his recognition that instrument mediations also *transform* vision, although he is more sensitive to the fact that it does so rather than (phenomenologically) how it does so. Taking a borderline case between naked and instrument-mediated vision, Hacking cites one such transformation of vision:

> Consider a device for low-flying jet planes . . . the vertical and horizontal scale are both of interest to the pilot who needs both to see a

> few hundred feet down and miles and miles away. The visual information is digitalized, processed, and cast on a head-up display on the windscreen. The distances are condensed and the altitude is expanded. Does the pilot see the terrain? Yes. Note that this case is not one in which the pilot could have seen the terrain by getting off the plane and taking a good look. There is no way to look at that much landscape without an instrument.[51]

This, in fact, is an excellent example of precisely the kind of phenomenological comparative analysis previously undertaken in *Technics and Praxis* regarding the ways instruments transform ordinary space and time differences but are also notable in their distinction from ordinary vision. Thus, I have no trouble agreeing with Hacking that the pilot *sees* the now instrument-transformed terrain. By contrast, Hacking may be seen to belong to the same trajectory as that of his phenomenological compatriots—as an instrumental realist. Moreover, he arrives at this position precisely within the same territory as his Euro-American counterparts, by way of re-emphasizing the role of experiment and instrument in the process of scientific discovery and praxis.

The second "mind-philosopher"—although highly revisionist—to arrive at a position of instrumental realism is Robert J. Ackermann. His 1985 *Data, Instruments, and Theory* was published only shortly after the 1983 appearances of Heelan's and Hacking's books. His position, as we shall see, places him very close to Heelan in the issue of a hermeneutics of a "data text" provided by instruments, and to each of the other instrumental realists noted as well.

One motivation revealed within *Data, Instruments, and Theory* is very traditional and in keeping with the programs of the old philosophy of science. Ackermann wishes to retain the sense of progress and the accumulation of knowledge thought to have been threatened by post-Kuhnian philosophy of science. Ackermann wants to claim that (a) science does, indeed, progress and thus accumulate knowledge, and (b) it is at least pragmatically nonrelativistic in this progression. The book's jacket proclaims:

> Robert J. Ackermann deals decisively with the problem of relativism that has plagued post-empiricist philosophy of science. Recognizing that theory and data are mediated by data domains (bordered data sets produced by scientific instruments), he argues that the use of instruments breaks the dependency of observation on theory and thus creates a reasoned basis for scientific objectivity.[52]

Here is a position very close to Heelan's, apparently arrived at independently and clearly via a different route.

Ackermann's book is more than an examination of what I call the embodiment of science in that—from a generally Popperian perspective—he also addresses the role of the institutionalization of science. However, it is with instrumentation and its embeddedness within praxis and perception that Ackermann may be seen to come closest to the previously outlined phenomenological positions.

If Ackermann, against Kuhn, remains motivated by the desire to preserve the notion of a nonrelativistic progress in knowledge, his means are those which entail the notion of an embodied science. Ackermann echoes the themes of our first look at Kuhn in two respects: First, he acknowledges that the directions of the new philosophy of science are such that neither form of Modern philosophy—rationalism nor empiricism—is adequate to account for the actual history of science, particularly in its revolutionary episodes. "Neither empiricism nor rationalism can satisfactorily explain sufficiently *rapid* theoretical changes in the sciences."[53] What Kuhn had pointed to was precisely this phenomenon of rapid change in science. However, Ackermann also recognizes that Kuhn did not explicitly lay such changes at the door of the *technology* of science, its instrumentation. Kuhn remains—for Ackermann—a vestigial rationalist: "In rationalism—including the variety offered by Kuhn, there is a fixed and determinate theoretical background that determines the value of data."[54] What is lacking in both Modern philosophy of science and in Kuhn is an appreciation of the crucial role of the *instrumentarium.* It was this oversight that Ackermann sees as one source of unwanted relativism in Kuhn. Paradigm shifts are more often than not related to the history of the instrument. "The succession of scientific ideas must be related to the succession of scientific instruments, and without such an underpinning in data the notion of shared paradigms and exemplars cannot be fully fleshed out. Kuhn's account misses the impact of new instrumentation and new data."[55]

If Kuhn and Foucault have become the interpreters of discontinuity in the progression of the sciences, both may be seen to fall prey to a temptation to see each paradigm or episteme as a finite system which has not evolved out of its predecessor and is doomed to eventual overthrow in the next paradigm or episteme. Rejecting this view, Ackermann announces that what breaks the connection between theory and observation is the *instrument:*

> The discussion introduced the view that scientific instruments break the connection between theory and observation, allowing the dialectic of theory and data to take place, and that the use of instruments establishes *data domains,* which are what theories adapt to. Instrumental splitting of theory and data means that data can be gathered independently of current theorizing, and used to constrain that theorizing. When new data domains are established by the use of new instruments, older data domains and the theories that are adapted to them do not disappear, but they can be split off

> as settled in principle. As instruments are improved, a succession of data domains can be used to define an objective direction of scientific progress.[56]

Ackermann thus reintroduces the notion of progress into science, although one can say that this progress is more than anything else the progress of *technology*—science's technology—instrumentation:

> It will be argued that the history of instrumentation provides an undirectional explanation of progress, in that later, more refined instruments are uniformly preferable to earlier instruments directed toward obtaining data in the same domain, and that this fact is essential to understanding the creation of what will be called data domains for scientific theory.[57]

Similarly, what is accumulative is also directly related to the history and progress of science's technology: "Underlying the history of theory is the history of data text. What is cumulative in the history of science is the gradual refinement of scientific instruments once they are introduced until they produce data that seem to be robust in the fact of further refinement."[58] In turn, this statement can be read as the history of a refinement of instrumentation sufficient to establish the manifestness of the phenomena read through the data.

Ackermann's position, then, is within the philosophy of science which most clearly points to the need for a strong counterpart philosophy of technology. Indeed, from Ackermann's perspective, progress in science is not possible without technological progress which may or may not be independent from science. However, this theme is not taken up in *Data, Instruments, and Theory.*

Ackermann does pursue the notion that instruments produce a "data domain" which, in turn, is to be hermeneutically interpreted. And here he independently comes very close in the larger issues to the position previously taken by Heelan. A "data domain" is explicitly recognized by Ackermann to be a text-analogue which calls for a hermeneutic process: "Facts appear only as data text requiring interpretation."[59]

As with Hacking, there is little "perception talk" in Ackermann's account, although the relationship of the instrumentarium to human sensory abilities is recognized. More concerned with the adequacy of our discourses as descriptions of the world, Ackermann notes, "Our discourses describe an aspect of the world, and they describe the world most adequately at the rough level of our discourses, the level of human perception, not at the level of the very large or the very small with respect to that level."[60] Obviously it is both the micro- and macrolevels which can be expanded through instruments. Over and over again, it is through the instrument that the new and unexpected is discovered and brought into perceivability. Ackermann acknowledges this as an aspect of the history of science from its very beginnings: "Galileo himself regarded the telescope as adding a new and superior sense to man's panoply of natural and common sense. The telescope thus marks a sharp break in knowledge."[61] Similarly, at the microlevel, "Roentgen did not expect to see anything unusual when he developed the famous photographic plate, but he did. The very fact that language has a loose fit to the world allows such primitive observations of the unanticipated to be described, at least in graphic terms."[62]

If it is clearly from the newly appearing within or through the instrumentarium that science can progress, Ackermann also briefly recognizes that a perceptual modelling may also operate within science in a deeper and more regular perceptualist guise than most philosophers focusing upon propositional thinking may acknowledge. Ackermann's discussion is worth a longish quotation:

> In addition to the instrumentarium, philosophers have overlooked the importance of technological developments in the visual presentation of both instrumental design and factual material. Manufacture of scientific instruments would not be possible without exploded views of assembly. . . . Visual thinking and visual metaphors have undoubtedly influenced scientific theorizing and even the notation of scientific fact, a point likely to be lost on philosophers who regard the products of science as a body of statements, even of things. Could modern scientific world be at its current peak of development without visual presentations and reproductions of photographs, x-rays, chromatographs, and so forth? . . . The answer seems clearly in the negative.[63]

What science does, both perceptually and linguistically, is refine our discriminations. "The instruments of science can best be seen as refining and extending human sensory apparatus, and scientific languages as refining and extending the discriminations that can be coded into ordinary language."[64]

Ackermann thus recognizes, at least implicitly, a role for what I have called a perceptualist model operating within science, a model which is continued and extended through its embodiment in instrumentation. However, like Heelan, Ackermann's is a *hermeneutic account* of science's perceptions. But closer in certain respects to Hacking, Ackermann remains quite cautionary about the relation between interpretation and nature itself—the instrument is not seen to be as directly revealing of entities or of nature as is emphasized in Heelan's more optimistic account. "It will be argued here that the features of the world revealed to experiment cannot be philosophically proven to be revealing of the world's real properties, but that experiment produces a text of data that must be interpreted, and whose augmentation may not seem initially consistent."[65] The reasons

Ackermann gives for this cautionary note—apart from an implicit epistemology which does not want to develop a totally hermeneutic approach to natural reality, hence retaining a vestigial empiricist strand—are similar to those given by Hacking. Perceptions can be highly ambiguous, particularly at the early stages of discovery and instrument development.

> As we have repeatedly observed, the data gathered at the growing edge of science are confused and uncertain. It is exactly in this area that theory can be seen to guide perception. Bacteria, investigated by the early microscopes, were at the limits of resolution, and could be seen as having almost any shape. Saturn and Mars, investigated by the early telescope, were of ambiguous configuration, and could be interpreted by different investigators as having divergent features.[66]

However, this reservation shared by all instrumental realists points to a stage of instrument development prior to any maximally attained transparency and to early stages in newly appearing "data texts." Ackermann's own account points to eventually greater gains attained through instrument development and its role within science: First, the instrument plays the role of a constant, or invariant, in observation. "The advantages of a scientific instrument are that it cannot change theories. . . . Instruments create an invariant relationship between their operations and the world, at least when we abstract from the experience involved in their correct use."[67] Ackermann sees this invariance which at certain states can accommodate several, even contradictory, theories as guaranteeing a fact domain: "Instruments and techniques assure scientists that they are talking about the same thing, that is, some scientific fact, when they disagree."[68] This invariance is laid to the very notion of a technology and its operation: "The instrument is usually a machine. Properly used, it stays the same from experiment to experiment, and its reduction of data complexity to a common signal, like a number expressing a reading, tends to neutralize what may be a highly variable scientific temperament from experiment to experiment."[69]

A second reason for a cautionary approach to instrument-revealed natural reality is related to the very notion of progress itself. The data domain is constantly expanding because of instrument development.

Thus, one could say that the frontier areas where ambiguity can occur get larger, rather than smaller. "The total text to be interpreted, the text of data, is not fixed before interpretation. The text of fact is constantly expanding, and can seemingly be endlessly expanded. Thus the clash of interpretations of fact, the clash of theory, is always potentially resolvable by expanding the factual text."[70] Yet as the data domain expands, in its very expansion—to which should be added refined resolutions in observations through refined instruments—one can point to areas of strong agreement. "Controversy in science is not rational after sufficient text from reality has been produced so that only one plausible interpretation of it is known. Because text can be generated, the possibility of settling scientific controversy always exists, at least in principle, even if the techniques for producing crucial text are not always known in the early stages of controversy."[71] In spite of Ackermann's demur regarding reality-through-instrument-produced text, he does recognize what amounts to a permanent recognition of some such domains:

> For some instruments, for example, the microscope and telescope, no theoretical limits to such refinement seem imminent. But certain objects studied by these instruments have remained in the data base since their discovery, and information about them has been gradually more precise. A changing data environment for scientific theories is like a changing environment for biological species. Progress is not guaranteed, but theory contains adaptive measures that allow it flexibility in the face of such change.[72]

In the end, then, through a process of slowly refined and growing knowledge, Ackermann allows for well-defined and resolved areas of reality-through-instrument-produced data text. His is a variant upon a hermeneutic theory of perception.

Another feature marking instrumental realism in this context is the notion of the inextricability of science as embodied through technology. Ackermann is decisive on this point and calls for a reconceptualization of science with respect to technology. Like his peers, Ackermann recognizes that the tendency to regard science in purely theoretical (and propositional) terms is a particular temptation within philosophy. "The philosophical attempt to distinguish pure and applied science is perhaps also related to a familiar philosophical predilection to favor activities similar to philosophical activities."[73] Rejecting this predilection, Ackermann holds:

> . . . that the decision to put a wedge between science and technology, no matter what its other merits, has precluded an understanding of the dynamics of modern science, since modern scientific progress depends at least partly on the technology required to produce the scientific instruments capable of wresting data from reality, and is inconceivable apart from that technology.[74]

He concludes even more strongly:

> . . . Whether science and technology can be made conceptually distinct is somewhat secondary to the fact that modern science contains a major

> technological component in its instrumentarium. Modern science is inconceivable without accurate pictorial representation and without the highly specialized scientific instruments that scientists use to investigate nature. This fact makes the question of the independence of science and technology moot, even though those concerned to defend pure science have frequently not noticed this important fact. It would be possible to write the history of science in terms of the instruments that have been available for scientific use.[75]

Here, again, is a variant upon the insight that contemporary science is *technologically embodied science.* Its relation to technology is not simply accidental, but essential. And such a reconception of science is necessary for any adequate *philosophy of science.*

INSTRUMENTS AS INTERFACE

What this exposition has shown is that there is now at least a minor consensus among philosophers of science from both Anglo-American and Euro-American traditions that any adequate philosophy of science must consider seriously the role of instrumentation. Instrumentation is one aspect of science's essential embodiment. It is interesting to note that this consensus was apparently arrived at by different and largely independent motivational routes. The very lack of citation between the two strands of this instrument-centered set of issues is indicative.

There is a history of citation within the Euro-American side of consensus. As colleagues, there is inter-citation of some frequency between Heelan and myself, and we both have cited Dreyfus. None of the Euro-American philosophers, however, are cited by either Hacking or Ackermann, although the former's work preceded in most cases the latter's publications. Nor do Hacking and Ackermann cite each other. Thus, assuming ignorance rather than other causes, I would say that the arrival at the crucial focus upon instrumentation was a matter of parallel philosophical evolution.

I shall dwell in greater detail upon this consensus and its implications in the next chapter; however, the fact that there is such a minoritarian consensus and that it remains to date minoritarian should be noted here. Each of the five philosophers dealt with in this chapter clearly belongs to the traditions of what I have called the *new* philosophy of science, not necessarily in its Kuhnian senses but in the sense that they have recognized and expressed explicit dissatisfaction with any philosophy of science which remains restricted to the logical, linguistic, or proposition analysis of science alone. Rather, science must be seen in its more concrete sense, which is to say that its practices (praxis) must be examined as well as its "theory."

There is more: I have argued that part of that recognition of the broader and more concrete analysis of science entails its necessary embodiment in technologies but that it also implies at least indirectly and, better, directly, consideration of both *perception and praxis.* These factors have shown themselves in the different variations upon the same themes in each of the five thinkers cited here.

Since this is probably the first time all have appeared lumped together in this way in print, making it particularly doubtful that each individual would easily accept the nomenclature, I shall dub this group of philosophers the "school" of *instrumental realists.* From the perspective of the philosophy of science, this group of philosophers has raised to high importance the consideration and need for both analysis and critique, the role of instruments in delivering both new knowledge, and a mediated mode of access to the world, particularly in its macro- and micro-aspects. The variations upon how, how much, and in what way these results are attained themselves demonstrate the need for further debate and consideration. But these instrumental realists stand in considerable contrast to the older and more abstract considerations of science as a "pure" theorizing activity.

Science's technology—focused in instrumentation but not limited to that—is also the *interface* with what might rightly be called philosophy of technology. While the philosophy of technology necessarily extends beyond philosophy of science in any number of directions, in its contemporary sense it cannot avoid the overlap with philosophy of science suggested here. Indeed, my own history and motivation took that direction in the mid-seventies and continues from within the recognized boundaries of the philosophy of technology (I do not regard myself primarily as a philosopher of science as such, even though I have placed myself in that context for purposes here).

V. Instrumental Realism

At the beginning of this itinerary, I pointed to the different and often divergent interests, concerns, parentages, and traditions associated with two of philosophy's subdisciplines: the philosophy of science and the philosophy of technology. With the appearance of what I call loosely the "school" of instrumental realists, I have located an important area of interface between these two subdisciplines. Of course, this group of philosophers remains a minority within both of the subdisciplines which continue to be dominated by the habits and concerns of earlier momenta. There is also something of a favored selectivity to this group of thinkers—either they are individuals who have always been recognized as philosophers of science, or if the interest arises out of philosophy of technology, as in my case, it is because their interests are focused in counterpart epistemological and ontological issues.

Yet the very convergence and apparent consensus regarding a critique of dominant momenta and what is shown in the positive analysis of experiment and instrument as the location of one technological embodiment of science is itself of considerable importance. I shall then trace both the broader areas of consensus which have appeared between philosophy of science and philosophy of technology, and also between the Anglo- and Euro-American traditions within which this convergence has occurred. Again, as in all the previous discussions, I will remain somewhat more on the side of the implications for philosophy of science as seen *from the contact with a concern for technology* than from what remains often more central for philosophers of technology.

First, the consensus regarding a critique of the extant and dominant forms of philosophy of science: I have already pointed out that particularly in North America, the dominant strands of philosophy of science arise from positivist and later, analytic roots. This has meant that the conception of science and the mode of analysis of science has largely been a matter of logical, linguistic, or propositional analysis. Of course, both the narrowness of that tradition and its tendency to disembody science from many of its concrete dimensions has come under attack—at least since the emergence of the new philosophy of science. Placing that attack in perspective and relating it to the underappreciated European developments which preceded or paralleled Kuhn was part of the program of the earlier chapters. I have tried to show, moreover, that much of the counterattack directed against "irrationalism" and "relativism" continues to miss the mark and that, rather, what the dominant philosophies of science have failed to note are the elements of praxis and perception which are operative in science. Part of what I take to be the consensus of the instrumental realists is either the direct or the implicit recognition of those elements of scientific activity.

Given the current state of the philosophy of science, with the virtual collapse of the older Positivist strains and the pluralization both through new concerns and the intrusions of sociological and historical revisions, it would be too strong to claim that what I have identified as the school of instrumental realists stands as unique and inflated within the now divergent directions of the *New* philosophies of science.

This is not even the case with respect to the emergence of what could be called *praxis* philosophies of science. A healthy development here has been the rise of the sociology of science to its current sophistication. Equally healthy is the emergence of a new generation of bilingual philosophers of science, trained in both Anglo- and Euro-American traditions, who break the older citation gaps earlier referred to. In the next chapter I shall return to this updating of both the moments I have chosen. But here, following the trajectory of praxis-perception philosophies, I wish to further cement the distinctiveness of the instrumental realists.

The focal point at which instrumental realism emerges is the simultaneous recognition of what I have called the *technological embodiment* of science, which occurs through the instruments and within experimental situations; and of the larger role of praxis and perception through such technologies. It is that focus which locates the positive side of the instrumental realist consensus and which is ultimately more important than the negative critique of extant philosophy of science. Nevertheless, the emergence of a critical consensus from such different starting points of the negative critique is also initially informative.

In its broadest sense, the five philosophers examined here all are conscious *and critical of* philosophy's predilection for the "purely theoretical," often with an accompanying disdain for, or ignorance of, praxis. In general, "armchair" philosophizing is seen as removed from the actual activity of the subject-domain which is being investigated. This predilection is a selective handicap when it comes to either appreciating, or even noticing, the practices of actual science. In this, Dreyfus is the most pointed in critique. He lays the predilection primarily at the door of one strand of tradition—the Platonic-Cartesian—which he thinks has become dominant in science at the expense of other possible traditions such as the Aristotelian through phenomenological ones. But in all five cases and beyond the most general recognition of the predilection, a much more specific critique emerges: *Any philosophy of science which is limited solely to linguistic, logical, or propositional methods* will not be able to adequately account for large sectors of scientific activity which entails more than such "rational" procedures alone.

The occasions for this critique—which may be seen as a general challenge to much older philosophy of science—vary among the members of this "school." Hacking, clearly hostile to older positivism, also wishes to overcome the penchant for "semantic ascent," which wants only talk about talk concerning such activities as observation and calls for a return to the analysis *of* observation; I take this to be an Anglo-American equivalent to the Husserlian "things themselves." Ackermann criticizes the same penchant because it causes dominant philosophy of science to overlook the dynamics of science, thus causing it to fall prey to what he continues to regard as the irrationalist-relativist aspects of the new philosophy of science.

The three phenomenologically oriented philosophers arrive at the same point, although more out of agreement within their own traditions about the role of perception and the embodied position of all epistemological activities noted in phenomenological epistemology. The primacy of praxis and perception is much more explicit within this tradition than in those of the Anglo-Americans.

Beyond the recognition of a general predilection toward theory and the overlooking of praxis, and beyond the general critique of methods linked solely to propositional dimensions of analysis, there is a third aspect within the negative consensus which also may be noted. This aspect arises out of certain results of the positive analysis of the instrumental. I anticipate that result here. By turning to the role of instruments and taking account of what they deliver for science, the very territories previously taken as theoretical domains also change—they shrink in size and significance. In short, instrumental realism gives some degree or type of "reality-status" to entities often taken to be merely theoretical, leaving only small areas to remain theoretical. This means, in turn, that the role of "pure" theorizing gets reduced to an even smaller area of science's activity than had previously been assumed. Theorizing becomes a special, highly speculative exercise of scientific imagination—important, but both reduced in size and open to greater skepticism—in regions outside the current reaches of instrumental possibility. Interestingly, this indirectly implies a more positive evaluation of the results of science than has been the case in either of the parent traditions.

Here, then, we have an emergent consensus regarding a critique of much extant philosophy of science. This critique is arrived at differently in the two traditions, as would be expected within philosophy: At one extreme we find Dreyfus, who regards a whole strand of tradition to be in error, even considering what Hacking would call the representations of science; but Hacking also arrives at a similar point through his appreciation of a more Baconian science, which both *intervenes* and represents, and who, through this vision concerning science, wishes to restore a balance between science's theoretical-representative activity and its intervening activity in experiment. From a slightly different angle, Ackermann wishes to restore the much older faith in *progress* in scientific knowledge, which he takes to have been under attack since Kuhn. But in the process, he must also attack the narrowness of the tradition which overlooks the dynamics of an instrumentally oriented science. Heelan, perhaps the most optimistic about what instrumentation can deliver, and I, more cautious, find the same results from the way in which instruments embody perception.

The negative critique also has a carry-over to the previous traditions from which both Anglo- and Euro-American thinkers have progressed. If the narrowness and shortsightedness of positivist-analytic philosophy of science, insofar as it restricts itself to propositional analysis, is an object of attack, there are criticisms directed at more direct parentages as well. Such critiques of their own respective traditions cannot be seen except, in part, from the positive results of a newly attained stance. And that stance is taken from the appreciation and analysis of the positive role of instruments within science. I shall shortly turn to that role, but in transition, we may take account of how this shift to the focus upon science's technologies has effected a critical role within the relevant philosophical traditions.

In each case, the critique carries with it a reevaluation of science itself. From the side of the Anglo-Americans, it is clear that Hacking is restoring something of a balance to what had been a serious imbalance of philosophical concerns. By reestablishing his more Baconian history of science, Hacking calls for and illustrates a more balanced view of science as *experimental,* and, in that process, necessarily deals with science's technologies. This reevaluation finds particular result in the reappraisal of observation, which not only becomes a direct concern for Hacking (rather than located through semantic ascent) but is re-placed in its sensory and often instrument-mediated context. In the process, Hacking denies any sharp distinction between "observation" and "theory," part of the critique of which opens the way to a positive role for instruments. They not only make phenomena present (or *manipulable,* the favored notion of Hacking) but go on to enhance sensing. Like each of the others, he also recognizes that a "theory bias" frequently distorts the history of science. Hacking then moves away from both a propositional analysis alone and a sharp theory/observation distinction toward the analysis of the role of instrumentation, which is usually overlooked entirely or underplayed in the dominant tradition.

Ackermann makes a somewhat stronger claim for instrumentation by linking the progress of science itself to the progress of instrumentation. Its history needs rewriting in this light. For Ackermann, the usual distinctions between science and technology need reexamination, and his is clearly an appreciation of even larger aspects of science's embodiment in technologies. Ackermann comes from a somewhat Popperian heritage which is also strong on the social institutionalization of science, and is itself abstracted out of the dominant traditions.

Ackermann recognizes that in a way parallel to what I shall momentarily point up in the European traditions, even the new philosophy of science lacks appreciation and insight into the role of instrumentation in scientific discovery. Kuhn is criticized by both Ackermann and Hacking for remaining too theory bound and too unappreciative of instruments in "paradigm shifts."

The same lack of concern for technologies appears within the European traditions. Indeed, the lack of focus upon a technologically embodied science lies at the root of two often noted problems in classical phenomenology. What appears as the wider attack upon theory/observation distinctions also plays a related role in the Husserlian traditions. In their earliest form, these traditions distinguish between the bodily-perceptual aspects of a lifeworld, a position which strongly differs from the "world" of science with its "theoretical" or "ideal" entities. The latter are presumably *reduced* entities in the Modern philosophical sense in having been reduced to "geometrical-physical" entities lacking sensory dimensions, and they are always derived from the plenary, sensory world of bodily existence.

Heelan points out that Wilfred Sellars has developed a similar distinction (drawn in part from Sellars's reading of Husserl); but in Sellars's case, there is an inversion of what is taken to be foundational. Whereas Husserl regards the *primary* world to be that of bodily-sensoriness, which becomes the "manifest" image in Sellars's inversion, the "scientific" image, which is the microworld of physics, etc., becomes primary and the base for Sellars's explanation. I have also pointed out that with an almost, although re-inverted, Sellarsian distinction, Merleau-Ponty continued the Husserlian version of primacy in what he calls the "distance" between the world of science and the lifeworld.

The earlier position of classical phenomenology did make a distinction between these domains in terms of a theory of *constitution.* Ordinary objects in comparison to scientific objects are *differently constituted.* But what the classical tradition did not note was that the latter are often, if not typically, *instrumentally* constituted. Technology—

instrumentation—makes the difference. Lacking the appreciation of the role of instruments which could embody perception, the earlier phenomenological tradition tended to interpret science's constructs as derivative and abstract, i.e., one could say "merely theoretical," in comparison to the plenary richness of primary perception. Given this implicit evaluation concerning the comparative reality-status of entities, it was often implied that phenomenology remained negatively predisposed to science.

The insertion of the instrument in a role seen to embody and enhance precisely the bodily ("material") perceptual intelligence recognized as central in phenomenology was to close this gap between the lifeworld and the world of science. And with it, the residual "negative" characterization of science also could disappear. Here then is a critique of classical phenomenology from the perspective of a more contemporary Euro-American phenomenology, focused specifically on a new appreciation of instrumentation as the embodied mediation needed to make the unification of lifeworld and scientific objects possible. Thus, from within both the Anglo-American and the Euro-American traditions, the instrumental realist consensus arrives at the interface between science and technology—and between the philosophies of science and of technology.

If the critique's negative dimension is important, this is more in relation to problems within related traditions and to the possibility of reformulations of problems which arise from such perspectives than it is to the more important aspects of the *positive* dimension of instrumental realism. Thus, I shall now turn to the features of that positive role.

In its broadest sense, the instrumental realist consensus points up the importance of science's technologies as the means by which discovery occurs and knowledge is expanded. I put the issue in this common sensical form first, even if it is a weaker claim than the one I think more interesting and penetrating, i.e., that contemporary science is more than accidentally—it is *essentially*—embodied technologically in its instrumentation.

Again, as we might suspect, there is a spectrum of opinion on the role of this technology, albeit focused on the role of knowledge-gathering instrumentation. Hacking's approach is again the most common sensical and conservative. His argument for the need to focus upon instrumentation seems to be largely historical-empirical. Science has always been Baconian in that Modern science not only represents, but intervenes. Empirically, both representation and intervention are—at least sometimes, and now, increasingly—tied to instruments or technologies. Not to regard this fact is to ignore important aspects of how science actually operates.

Hacking's call for a re-Baconized concept of science, however, retains its concern for a positive role for science's representations as well. What Hacking does, in part, is to say that if philosophers of science want to continue some form of realism, they have been looking in the wrong direction. They would be better off seeing that experimentalists become pragmatic realists in the familiarization of interactions occurring through instrumental manipulations of the entities which become part of the furniture of a scientific world.

What complicates making too strong a claim about an essential role for instrumentation in contemporary science is Hacking's "pluralism" of methods. He believes in both a plurality of scientific methods and a plurality of experimental approaches. Also, with respect to another issue which divides our five philosophers, he clearly does not hold that *all* scientific observation is either tied to or occurs exclusively in the presence of instruments.

Yet a closer examination of Hacking's examples and more particularly, his claims about what kind of activity observation presupposes, reveals that he is at least implicitly much closer to the harder positions taken by others of this "school." It is abundantly clear—and I shall not cite the many relevant passages which support this—that observation in its scientific sense (a) is a skill, (b) is a matter of trained learning, (c) is active and not passive seeing, and (d) sometimes contains exceptional insight peculiar to technical skills.

Moreover, what the examples show is that (1) almost all observations occur in an instrumental context, (2) there is an implied human-instrument relation, and (3) even what is cited as "sense data" reporting is instrument-mediated and/or constituted. A particularly telling example is his citing of William Herschel's noticing a phenomenological difference between heat and light transmission:

> He had been using coloured filters in one of his telescopes. He noticed that filters of different colors transmit different amounts of heat: "When I used some of them I felt a sensation of heat, though I had but little light, while others gave me much light with scarce any sensation of heat." We shall not find a better sense-datum report than this, in the whole history of science.[1]

Herschel went on to use thermometers to test the same phenomenon, finding thereby that "not only orange warms more than indigo, but that there is also a heating effect below the visible red spectrum."[2] Herschel then surmised that there were both visible and invisible rays coming from the sun—only much later was this observation appreciated and understood theoretically.

Reverting to my own terminology, I would note that (1) here is an observation thoroughly technologically contexted, (2) the first observation of heat and light differences "sensed" by Herschel were (a) instrumentally mediated and (b) of what I call the embodiment sort, i.e., perceptually felt. The second set of observations, which used thermometers while testing the "same" phenomenon, were of the hermeneutic sort, both mediated, and since "read," of a more indirect type than the first set of observations. But both were clearly instrumentally contexted and mediated—and in the stronger sense noted by Heelan, instrumentally "constituted." The effects were "carpentered" via the prism and lenses.

The strong sense of a technologically constituted data-domain, while clearly pointed up by Hacking, is not made with the same force as it is in Ackermann's, Heelan's, or my accounts. Yet there is the recognition that many such areas of experimental results are a growing field of observation. These include all those micro- and macrophenomena which lie below or beyond ordinary sensory capacity and which are made present only through or with instruments. At the least, the instrument serves as a condition for such observations; Hacking recognizes and analyzes these conditions in his history of the microscope.

If Hacking is the most cautionary of the philosophers with respect to the essential embodiment of science in technology, Heelan makes the strongest—if also the narrowest—claim to the essentiality of this form of embodiment. For Heelan, as we have noted, *only* those phenomena which have been instrumentally "carpentered" and "constituted" can have claim to scientific "reality." This is in part because, for Heelan, all scientific perception is simultaneously perception plus *measurement;* for measurement to be scientific, it must entail a standardized technology or instrument. Here then is an approach which holds to a *necessary connection* between scientific observation and its technologies. Of course, one outcome of this position is also that in contrast to Hacking's pluralism of methods, Heelan comes close to holding that there is *a Scientific* method.

While such a strong position might appear to narrow what can be taken as the "world" of science because Heelan is also the most optimistic about what realities are exhibited by technologically embodied science, that is not the case. For Heelan, the World is open to both a pre-scientific and a scientific mode of perception. The pre-scientific World is clearly limited to noninstrumentally mediated perceptions, while the scientific world is perceived only through instruments. Adapting a version of Sellars's "manifest"/"scientific" image distinction, Heelan claims: "A scientific image . . . represents objects as constituted in their essential forms by systems of postulated (or theoretical) entities, related to one another and to some manifest World by scientific theory, *and encountered only through the mediation of instruments or technology.* [Italics mine]"[3]

What makes the pre-scientific and the scientific worlds overlap is that both are "perceived"; however, the latter is perceived only through "readable technologies." Heelan argues that such a "reading" is *direct* and equivalent to perception. It is, however, a technology-embedded perception through the "readable technologies" of instrumentation. In this, Heelan continues the physicist's use of such observations as "direct observation," a use which Hacking criticizes."[4]

Although, from my point of view, Heelan collapses what I distinguish as a difference between embodiment and hermeneutic relations through technologies, his hermeneutic account of "readable technologies" does make of science a specialized mode of world exhibition which, on one side, also "perceives" the world, although only through both formal and technological mediations. Its range of phenomena to be revealed both overlaps the directly perceived world and enhances it, insofar as any entity which is instrumentally detectable becomes a candidate for a reality-status.

Dreyfus occupies a peculiar status within the spectrum of instrumental realists regarding the relation between science and its technological embodiment. That is largely due to the fact that his program is very different from the overlapping ones of the other four philosophers being discussed. On one side, it might seem that Dreyfus is the most *negative* of the five with respect to the role of technological embodiment of science, and also the closest to the older phenomenological tradition's suspicion of the role of science. But that is partly because Dreyfus is not directing himself to instrumentation as such, but to a certain interpretation about a special use of machines. His program strongly *contrasts* machine and human perceptual results. Only indirectly, so far as the instrumental realism at issue here, can the notion of technological embodiment be noted in Dreyfus. But it is there: A subcontention might well be that AI is the attempt to embody a precisely Platonic-Cartesian theory instrumentally (and since this theory is wrong for Dreyfus, it must fail as a model for human embodied intelligence).

Ackermann, too, occupies a fairly strong embodiment position in the science-technology interrelationship. Interestingly, he argues against one traditional "pure theory"/"applied science" argument on the grounds that there may be an ulterior motive to preserving such a distinction by making "pure" theoreticians less open to moral, social, or political criticism (pure knowledge presumably is neutral). Ultimately, Ackermann argues that such a distinction between science and technology has "precluded an understanding of the dynamics of modern science" and that understanding science "is inconceivable apart from that [instrumental] technology."[5] Short of Heelan, however, Ackermann sees the dependence of science upon technology as only partial. Technology, nevertheless, provides a major component of modern science in its instrumentarium. He does not, however, argue that only instrumentally constituted entities count as scientific entities.

My position is close to Ackermann's in terms of placement on the spectrum. I argue that science, in its modern sense, has differed from classical (Greek) science from the beginning in its technological embodiment, both with respect to the actual use of instruments, as in the case of Galileo. His self-conscious use of instruments was proclaimed in "The Heavenly Messenger," one of the first science newsletters. But such use of technologies was also already part of the Renaissance lifeworld of the time. Modern science is and has been essentially and historically technologically embodied.

Of course, each of our authors notes the obvious: that science is now increasingly so embodied, and necessarily so, with respect to those regions of investigation which fall below or above what is directly sensed in unaided perception.

With respect to what I call a technological embodiment, there remains something of a spectrum of opinion which also relates to the third feature of the consensus binding the instrumental realists together. This, too, is a strongly positive feature of the consensus. For whether science constitutes its entities either partially or solely through technological constructions, one one leaves the realm of what is visible to unaided perception in the micro- and macrolevels discovered by science, all clearly agree that enhanced perception occurs only through such instrumentation.

If we momentarily disregard the differences of opinion about whether or not all perceptions of science must be instrument related—within the realm of the micro- and the macrolevels now occupying far larger regions in scientific interest than the "middle-sized" ones of unaided perception—the question of an essential embodiment becomes a matter of agreement. As I noted previously, Heidegger argues that "technology reveals a world," and here we find a new sense for that phrase—technology reveals the micro- and macroworld which lies beyond unaided sense.

Our instrumental realists emphasize this enhancement and magnification made possible through instrumentation. It is here that the narrowing of the region of what has previously been thought of as "theoretical" becomes replaced with the instrumentally "observable," and in differing degrees, this observability in turn becomes part of a new *perceptual* region. Here is the heart of the "realism" of instrumental realism. But here too are the areas of most interesting variations upon the broader theme.

On the spectrum of a reality-status for instrumentally delivered entities, Ackermann takes the most conservative position. He argues both for the largest degree of ambiguity relating to what he calls "data domains" which are text-like, and for considerable skepticism relating to the (hermeneutic) interpretive process. "Instrumental means only produce a data text whose relationship to nature is problematic."[6] And, "the features of the world revealed to experiment cannot be philosophically proven to be revealing of the world's properties."[7] (I am not sure whether Ackermann holds any counterpart thesis that other means of analyzing world properties *can be* philosophically proven!) Yet, examination of the *interpretive result* relating to such data domains turns out to be the same as for other knowledge claims.

The facts that knowledge progresses and accumulates through instrumental progress, that instruments serve as invariants for multiple observations, and that refinement leads eventually to sufficient agreement all point to the strongest claim Ackermann can make within his dialogical and consensus-driven interpretation of (hermeneutic) science. At some point, further argument—and by extension, skepticism of an extreme kind—become irrelevant.

At the other end of the spectrum, but only in a nuanced way, one again finds Heelan. Within his narrower definition of scientific realities, the instrument delivers most. This is not at all to say that Heelan disregards the ambiguities pointed to by each of the other thinkers in the discussion of instrument-mediated phenomena. He discusses at length precisely the problem of "textual" opacity, ambiguity, etc., which worries the other instrumental realists. But he remains the most optimistic in relation to what is taken to be the "directness" of instrumental mediation and to what I term the collapse of perceptual characteristics into hermeneutic ones. "Reading" becomes a variant upon (direct) perception. It will be seen that both Ackermann and Heelan emphasize a mode of hermeneutic "seeing," despite the fact that they respectively occupy the two extreme positions on the spectrum of reality-deliverance.

By contrast, Hacking and I occupy the middle positions of this spectrum, with my position somewhat more optimistic than Hacking's regarding reality-deliverance. Our midpositions are qualified due to concerns over not only *what* is delivered but the way in which the instrument *transforms* as well as delivers the phenomenon in question. However, we arrive independently at our concerns for instrumental transformation of phenomena, with Hacking taking what I shall call an externalist approach (inferring what can be seen via current physical theory, as in the application of optical theory to visual result), whereas I take a phenomenological approach utilizing variation and difference from the multiple profiles which occur between instrumental and non-instrumental contexts.

In spite of these differences, there remains consensus regarding the positive delivery of scientifically constituted entities, the role of instrumentation in making this possible, the enhancement of perceptual role, and the subsequent narrowing of the area of "theory" previously thought to lie outside what is (scientifically) experienceable. Yet our four philosophy of science-oriented thinkers all point to a significant and agreed-upon realm in which there are instrumentally delivered realities. These realities need further attention, but they also constitute a large and interesting realm at the interface of science and of technology.

If the technological embodiment of science becomes the focus for what can be analyzed within this interface area, it is in the domain of praxis-perception taken broadly, but also somewhat thinly, which becomes the last part of the instrumental realist consensus. The *praxis* part of that model is clearly the stronger element within the consensus. This is because each of our philosophers has taken activity and particularly the instrument and experiment-oriented activity of science more seriously than older dominant forms of analysis. Furthermore, each recognizes the wider, more institutionally oriented ways in which science operates within contemporary Big Science—in a more thorough fashion than earlier positivistic or analytic modes of interpretation. Ackermann's position is the strongest in this respect, but in each case the others would find a related position. At the very least, both the instrument designer and the skilled users of instruments are raised to higher importance within the overall institution of science.

When it comes to the *perception* side of the praxis-perception model, I suggest that it might seem our set of philosophers would find the largest areas of disagreement. This may be expected because there is a clearer division here between the Anglo- and the Euro-American groups over what might count as a theory of perception. This is precisely because the latter come from a very strong and more sharply defined tradition regarding perception, in contrast to the vestigially empiricist theories of perception which might be expected of the Anglo-Americans. But that turns out not to be entirely the case. However, there is something of an implicit, contrasted-to-higher-degree explicitness involved here between these two groups.

Through these differences, however, there is also an area which could be called consensual. First, all five of our thinkers notice and agree that sensory-perceptual dimensions of observation are of high importance within the observational side of science. This agreement—other than being taken for granted but often overlooked precisely because so taken—might not appear exceptional, but linked to the more specific recognition of the role of *visual* imagination, modelling, and even metaphorization, it constitutes a strong sub-domain concerning the way science operates. This visualism is not only recognized in different ways by each of our authors but is held to be an essential part of scientific (perceptual) praxis. Even more strongly, there is mutual recognition that such visualism within science occurs in gestalt and often intuitional ways with respect to insight. What may be called "visual thinking" plays a much larger role than is recognized in most philosophy of science. And this visualism can and does play a role in and through instrumentation, as well as directly. I would like to argue that this part of the consensus indirectly calls for a careful *phenomenological* analysis of visual perception. That analysis is in part undertaken by each of the three instrumental realists of the Euro-American side.

A division among our instrumental realists is not provided by the above, but by two different ways of cutting into the role of observation in its perceptual dimensions. And these divisions do not follow Anglo-Euro-American orientations. The first such division relates to whether or not *all* scientific reality is constituted necessarily through instrumental mediation. At one level, only Heelan occupies an affirmative position on this issue. Each of the other philosophers holds to some variation of observation being possible, either through instrumental or through unmediated perception. While, common-sensically, Heelan's position may appear extreme, a closer examination shows that it is considerably more complicated, since such a position entails what is taken to be "science," and subordinately, what can be taken to be a "scientific perception."

All of our philosophers agree that any scientific perception is not simply an ordinary perception—and not because all scientific perceptions are simply "theory laden." Rather, our instrumental realists hold diverse views regarding the acute and active skill which must count as part of a scientific perception.

This is to say that there is a peculiar disciplining of perception which is required within science. But in addition to the implied hermeneutics of such a discipline, there are forms of *tacit* or unspoken skills which many times appear with skilled observers with (or without) instruments.

Hacking, for example, clearly holds that what I call a scientific observation need not be instrumentally mediated, although many are (and their numbers are increasing). But historically, many important observations occur both without instruments (unaided perception) and, more importantly, without an explicit relation to measurements. In contrast, Heelan's position holds that scientific perceptions are through "readable technologies" and are also perceptions *plus measurements.* Hacking, like Heelan, recognizes that in more recent science, there are many technologically devised experiments which produce their own carpentered results which, in turn, become highly important to science itself, even if the artifacts of those experiments are not found directly in nature. Indeed, the previously cited example of the laws of thermodynamics arising out of observations concerning the steam engine are interestingly reexamined in precisely that way by Hacking. It might seem that there is a deeper division here than turns out to be the case.

Hacking also points out that the obsession with measurement is historically a late moment in the history of modern science, although in a stronger sense it does characterize most contemporary science. Granting that measurements have been part of scientific praxis in some sense since even before modern science (among the Babylonians, for example), Hacking points out, "*Our* conception of numbers and measuring is clear and unquestioned only at the end of the nineteenth century. . . . Kuhn suggests that there was a second scientific revolution, during which a spectrum of physical science is, for the first time, 'mathematized.' He puts this somewhere between 1800 and 1850."[8] Given what I have called Heelan's implicit use of physics—one should now say *contemporary* physics—as the model for science, in contrast to Hacking's insistence on a pluralism of both theoretical and experimental "sciences," one sees that the argument is about what shall count as science itself.

However, the issue of necessary technological constitution does not relate simply to measurement. In the context of developing the notion of observation as a skill—often independent of and in advance of theories which can accommodate observations—Hacking notes that:

> Caroline Herschel (sister of William) discovered more comets than any other person in history. She got eight in a single year. Several things helped her do this. She was indefatigable. Every moment of cloudless night she was at her station. . . . But most important of all, she could recognize a comet at once. . . . [She] could tell a comet just by looking.[9]

This skilled, persistent, and gestalt form of scientific perception was not a matter of measurement per se, nor was it something inferred from mathematical projections—although it could have been. Hacking indicates, "Everyone except possibly brother William had to follow the path of the suspected comet before reaching any opinion on its nature." (Comets have parabolic trajectories.)[10]

Nor, interestingly, were these brilliant observations initially instrumentally mediated—most of her locating observations were made without the use of even a telescope—but on closer examination, they *did occur in an instrumental context* in a different way. "She used a device, reconstructed only in 1980 by Michael Hoskin, that enabled her, each night, to scan the entire sky, slice by slice, never skimping on any corner of the heavens."[11]

This would appear to be an example which is, at its core, both unmediated and non-measurement oriented, yet clearly a "scientific perception." Yet while the core of the example does illustrate this, there is also now a recognized instrument-context in both the scanning device and the later telescopic confirmation of initial observations. This example points to a related question, one which focuses more pointedly on the variant understandings of what count for perceptions within the domain of "scientific perceptions."

Heelan's notion of scientific perceptions is perhaps more narrow than that of the others, necessarily linked to instrumental contexts; but it is also cast in *hermeneutic* form. As I have earlier noted, his closest compatriot in this respect is Ackermann. Instruments provide scientific reality through their "readability" and a resultant "data domain," which is continuously interpreted. The difference between Ackermann and Heelan, at this point, is their understanding of a role of interpretation. Heelan's greater optimism concerning instrument delivery of scientific "reality" is related to his inclusion of the features of *direct* perception within his hermeneutic notion of perception.

Within this issue lies the second area of disagreement among our instrumental realists. If the first is to determine whether all, or only some, scientific perceptions are instrument mediated, the second is over the issue of how perception itself is to be construed. It is the issue which lies between a totally hermeneuticized and a body-retained perception.

The issue is not clearly divided between the Anglo- and the Euro-Americans, although it becomes sharper from within the Euro-American tradition than between the two traditions. But first, a brief look at the Anglo-Americans: As already noted, Ackermann's position is largely hermeneutic—instruments produce data domains which are interpreted. But interpretation is largely a matter of gradual refinement and arrival at a consensus within the institutional context of science. Ackermann, in short, does not often focus upon individualized perceptions. Also lacking is any keen sense of bodily position as an important factor in perception. Instead, he focuses upon what I have earlier called "macroperception," i.e., perception in its largely cultured sense. Such a perception, of course, is ipso facto a matter of larger hermeneutic concern. And because it is this institutionalized form of perception which plays the larger role in Ackermann, he overlaps Heelan's hermeneuticizing of perception.

In some contrast with Ackermann, Hacking does draw heavily from historical *individuals* and upon a version of sense-data tradition which implicitly related to individualized, hence embodied, observers. In that indirect way, Hacking can overlap from a neo-empiricist perspective the emphases of the bodily oriented phenomenologists. Thus, in some areas, Hacking comes closer to both Dreyfus and myself.

Stating the issue as one in which some strength of distinction is to be maintained between a bodily-perceptual (micro-) and a hermeneutic perceptual (macro-) mode, Dreyfus again emerges as having the strongest body-perceptual emphasis. Not only is the human body implied in all its intelligent actions, but its positionality, its gestalt features, and particularly its kinesthetic motility determine its perception of reality.

Not that these elements are ignored by Heelan, the most extreme of the hermeneutic perceptualists among the Euro-Americans. Indeed, his epistemology quite explicitly recognizes and elevates bodily positionality as a reason for the rejection of both transcendentalism and the implied all-seeing "eye" of Platonic or Cartesian interpreted science. Yet in the end, bodily perception is collapsed into hermeneutic perception. It is the text metaphor which becomes dominant in this version of scientific perception.

This hermeneuticizing of perception may be rephrased by noting that Heelan emphasizes what he takes to be the *direct* perceptual aspects of reading. " . . . The value of that quantity [of a scientific entity] can be 'read' directly from this response (providing one is experienced and skillful in the use of the instrument)."[12] Such experienced "readings" are, of course, perceptual experiences. There is a gestalt quality of any form of skilled and already learned reading. And because these perceptual characteristics also characterize the hermeneutic use of instruments, Heelan justifies the collapse of the perceptual aspect into its specialized form of "reading."

> "Reading" temperature is like reading a text . . . a "text" is "written" causally by the environment under standard circumstances . . . on the thermometer, and this "text" is "read" as being "about" a presented object, here, the temperature. . . . Such a process is, I claim, essentially both *hermeneutical and perceptual.*[13]

Yet, as all the examples show, it is the hermeneutic dimension which absorbs the perceptual one.

If Dreyfus's position clearly emphasizes the bodily perceptuality of thought and Heelan's position hermeneuticizes bodily perception, my position falls between these two. As noted before, I distinguish among several modes of human-technology relations which may be phenomenologically analyzed. Of these, *embodiment* relations and *hermeneutic* relations are the most relevant to this context. In this distinction, both how one experiences the instrument and the deliverance of the *referent* make a difference. I may illustrate this difference by drawing from both Hacking's Herschel siblings and Heelan's thermometer examples.

Caroline Herschel's eyeball observations of comets were *direct* perceptual observations. Clearly, within the context of a highly skilled seeing, and informed by background information on differences between comets and other eyeball astronomical phenomena, she was able to bodily perceive with gestalt instantaneity her selected referents. When an instrument delivers just such a perception, isomorphic though transformed in some way—usually by a reduction of qualities—I call the perception "embodied." Such is the case with William Herschel's heat perceptions, mediated through lenses and filters. He *felt* the radiant heat upon his hand—and as Hacking went on to describe, he made claims for "measured" felt differences to the thousandths of degrees, which could not have been confirmed by any extant instruments of the time. The phenomenon of heat—transformed and reduced from the fuller presence of focused light—was bodily, perceptually, felt.

By contrast, both Herschel's totally instrument-mediated and Heelan's "read" thermometer produce a "text" which *refers* differently. It does *refer*, thus it does properly belong to the realm of instrumental realism; but it refers in ways not isomorphic with bodily-sensory relations to objects. Its "textual quality" is presenced in numbers, alphabets, or some other "read" feature which is not isomorphic with that to which it refers. Such hermeneutic relations, however, may also include various manipulable factors sufficient to keep them close to Hacking's instrumental manipulability of otherwise unsensed entities. In that sense and despite differences, each of the positions remains within the limits noted for an instrumental realism such as has been outlined here.

I shall not further pursue what I take to be both the identifiability of a "school" of instrumental realists and the variations within this "school" which may be areas of fruitful and interesting debate and discussion. Were I to have a wish as an author, I would hope that readers would read, in parallel, the philosophers' books related to the issues here discussed. It would be my contention that any time such a diverse group of philosophers arrive at such a consensus, often independently and even in some isolation from one another, the issues evolved may well be important ones. The broadest of those issues in the present context centers on the need to more deeply concentrate upon the role of science's technology, its instrumental embodiment. And this is the interface area which connects philosophy of science and philosophy of technology.

VI. The Experiment

Two moments in the history of the philosophy of science as it interfaces with the philosophy of technology have now been traced. The first moment was located in the twin appearances of a New philosophy of science—through Kuhn with his paradigm shifts and Foucault and his epistemes. By backgrounding both with the European phenomenologists Husserl and Merleau-Ponty, I tried to show the deep role played by a praxis-perception mode of interpretation arising within new philosophy of science.

The second moment was the emergence of the "school" of instrumental realists, arising from both Anglo- and Euro-American philosophy of science and technology. For while the first moment switched the interpretation of science toward a gestalt model of operation, neither the Kuhn-Foucault pair nor their phenomenological forebears made the focus upon science's *instrumental embodiment* central in the way the instrumental realists do.

This second moment was backgrounded first by Heidegger, who inverted the usual understanding of science and technology by making technology prior to science, and who found in the praxis of tool use an embodiment model for understanding the generation of knowledge. And then, in 1972–85 series of books I chose for examination, it could be said that Dreyfus was the more recent background figure, turning a particular area of philosophy of science toward problems of instrumental embodiment.

Reading the history of the philosophy of science in such a way has the advantage of displaying a certain clarity—but, like all clarity, at a cost. The cost is a certain over-simplification and overlooking of possibly equally important directions taken since the rise of new philosophy of science. For example, I mentioned but did not deal deeply with the generation of historians of technology who had already also overturned the too simple notions of technology as applied science and the primacy of disembodied theory. The '70s were marked by much development in the history of both science and technology.[1]

Then in the '80s there began to appear a powerful movement alongside the equally proliferating pluralism in philosophy of science, a group of studies of science arising from the sociology of science. If the revisionist historians placed a certain emphasis upon "science owes more to the steam engine than the steam engine does to science," the sociologists of science began to see the knowledge products of science as "socially constructed." Does science "construct" its realities?

And, if the post-Kuhnian era in philosophy of science was marked by the retreat—even productive demise—of much Positivism or Logical-Empiricism, the up side was a proliferation of different approaches to science. From the revisionist but still related mainstream of philosophy of science there arose a series of debates about "realism" versus "anti-realism"—to which this book belongs from a quite different perspective, with its emphasis upon *instrumental realism.* These debates arose from the '70s on.

More recently, there also appeared what could be called a hybrid Anglo-Euro-American philosophy of science in the '80s, led by younger philosophers schooled in both traditions and bringing to the new philosophy of science a more *hermeneutic* interpretation. Here the background figures are primarily Heidegger, sometimes Gadamer, and especially Foucault, now in new guise as the founder of a discourse-praxis which takes form as power/knowledge.[2]

While this précis of related movements may help locate instrumental realism in the broader field of interpretations, I do not intend this primer to be either encyclopedic or a survey of the entire terrain. Yet, from these related movements one more step is required to both fulfill and confirm the initial movements of instrumental realism. That step requires that science's necessary embodiment in technology be raised to the level of complexity and corporate structure in which contemporary Big Science operates in the experiment, *particularly Big Instrument experiment.*

Here once again, in keeping the the tactic of choosing key books to signal the moment displayed, I will focus upon the very historically oriented *How Experiments End* (1987), written by the philosopher of science Peter Galison; and on the work of the paradigm sociologist of science Bruno Latour in his *Science in Action* (1987).

However, before re-focusing upon their new level of the technological embodiment of science in the experiment, it would do a disservice to our original instrumental realists were we not to acknowledge their own awareness and development with respect to the experiment and the more complex levels of science's embodiment. For all five of the thinkers examined are active philosophers hard at work on the next steps arising from the original insights of the 1972–85 books cited.

In such an update, Hubert Dreyfus again emerges as a pace-setter and background figure. While still engaged in the debates concerning computers and artificial intelligence—expanded to expert systems and a whole spectrum of issues—Dreyfus has simultaneously become both the guru of a generation of hybrid philosophers of science and the penultimate victor in one sector of the computer design battles.

After *What Computers Can't Do,* which was modelled upon the body-praxis notions of Merleau-Ponty with adaptations from Husserl and Heidegger, Dreyfus began to be more involved with interpreting Heidegger and then Foucault. With Paul Rabinow, he produced in 1982 one of the earliest of the Foucault interpretations, *Michel Foucault: Beyond Structuralism and Hermeneutics.* From this combination, often disseminated through NEH summer seminars, Dreyfus and company began to have a strong influence upon a hermeneutic direction in the philosophy of science, but a hermeneutic which had strong Foucaultian discourse praxis overtones and strong undercurrents toward many of the theses in the sociology of science.

The strongest result of philosophy of science, per se, is Joseph Rouse's *Knowledge and Power: Toward a Political Philosophy of Science* (1987). Here the key background figures are clearly Heidegger and Foucault, now turned toward science as enmeshed in the larger political context.

Within science—and here I mean primarily computer science—Dreyfus has both had a strong impact and has updated himself. If in the '70s Dreyfus was *negatively* setting many of the program challenges for computer program designers—"if Dreyfus says we can't do it, we've got to prove him wrong by doing it"—by the '80s both the difficulties of doing what Dreyfus said could not be done and the alternative "phenomenological" gestalt models suggested by him began to make marks within computer circles themselves.

A historical climax to this development occurred in the fall of 1989 at Berkeley with a Dreyfus-run "Applied Heidegger" conference attended by some 700 persons. To everyone's surprise, nearly 150 of those taking part were computer designers, users, and business people now convinced of some versions of a praxis-perception utility to computer design. Echoing Heidegger, Terry Winograd began to call his version of design "ontological design," and his *Understanding Computers and Cognition: A New Foundation for Design* (1985) clearly is backgrounded by Dreyfus.[3]

My *Technics and Praxis* was first to follow Dreyfus in the late 70s. The initial and somewhat sketchy "phenomenology of instrumentation" developed then has since been much more systematically undertaken in last year's *Technology and the Lifeworld* (1990, in this series). However, in that philosophy of technology context, it is not science alone which is examined. Indeed, I parallel scientific instrumentation with musical instrumentation in one section to show that the human-technology interface displays related characteristics in two very different sets of praxes. However, *Technology and the Lifeworld* is more concerned with setting both science's and other embodiment technologies in the larger context of cultures, and arguments are addressed to the cultural embeddedness of all technologies.

All this is transitional to the ascent toward complex Big Science experiment. And with that focus, from among our instrumental realists it is Hacking and Heelan who come to the fore. Indeed, both are now at work on books dealing explicitly with experiments, and both have published significant articles on the complex experiment. We shall return to Heelan and Hacking shortly in this new level of context.

In all that we have so far surveyed on the rise of instrumental realism, it still could be said that the examples used and the implicit context presupposed have focused upon concrete (and first-person) *observation,* where the role of perception remains relatively clear, but the examples most often used could still fit within the older and *simpler* science than the science which predominates now.

Joseph Rouse, in his synoptic approach to new philosophy of science, notes:

> One might say that the traditional philosophical model of the local site of research is the observatory. Scientists look out at the world to bring parts of it inside to observe them. Whatever manipulative activities they perform either are directed at their instruments or are attempts to *re*produce phenomena in a location and setting where they will be observable. The new empiricism leads us instead to take seriously the *labor*atory. Scientists produce phenomena: may of the things they study are not "natural" events but are very much the result of artifice.[4]

There is a sense that this remains the tone of many of the examples previously discussed among our instrumental realists.

But it is the laboratory with its "construction" of "carpentered" entities which becomes the norm of much Big Science. Here we reach a new level of complexity which is basically new even to science itself. And there are several dimensions to this new level of complexity:

1) In Big Science what was once the identifiable, single investigator or, at most, a small group of investigators, has become the large, corporate enterprise. This is evidenced not only by the multiplication of science personnel, but by the very way in which investigations are reported. Rare is the single-authored report or article. An example of this change has been recorded in *The Scientist:*

> That the incidence of multiple authorship has increased is indisputable; the percentage of papers published with multiple authors in *The New England Journal of Medicine* has increased from 1.5 percent in 1886 . . . to 96 percent in 1977. This is due in part to the changing nature of scientific inquiry.[5]

2) Instrumentation, too, has moved from simple, often hand-crafted devices to the megamachines of today, for example the CERN and Fermilab accelerators which lead today's physicists into new areas. In *How Experiments End,* Peter Galison has pointed out:

> Through the 1930's . . . most experimental work could be undertaken in rooms of a few hundred square feet, with moderate, furniture-sized equipment. But at Fermilab, where one of the high-energy experiments we will investigate took place, a herd of buffalo actually grazes in the thousand-odd acres surrounded by the main experimental ring, and individual detectors can cost millions or tens of millions of dollars.[6]

This movement to the large and complex, while most dramatic in physics, belongs in different ways to most of the sciences concerned with micro-phenomena.

3) This brings us to another aspect of contemporary complexity. A virtual empirical "rule" which seems to be followed is that the smaller the phenomenon, the larger and more complex the instrumental set. In *Technics and Praxis* (and again in *Technology and the Lifeworld*) I observed this relation of small phenomenon to large equipment in a phenomenological observation about instrumental non-neutrality. I argued that any interesting instrument selects and transforms its reference-object, and that in the case of many scientific instruments, that selectivity enhanced either the macro- or micro-aspects of the phenomena being investigated.

This, in turn, opens the way to an instrumental trajectory, which if followed, *inclines* investigations in certain, rather than other directions. Here is another more subtle variant upon the technology-leads-science perspective.

4) And while the focus on much of the experiment will be upon the artifactuality of experimentally "constructed" realities and their associated ambiguities, another observation concerning the type of instrumental production is also worth making.

I differentiate between the human-instrument relations which display in relatively isomorphic displays their perceptual or bodily engagements with the observer. These I call *embodiment relations.* But those instrument displays which display either through the thoroughly textualized form of digits, numbers, or other text-like phenomena, or those which display results from indirect or sub-perceptual phenomena (such as bubble chambers), are more hermeneutic in form and are—at least vis-à-vis embodiment results—often more ambiguous. In complex experimental equipment, such latter types of display are often the more common.

Thus, there is a multiple complexity which arises in Big Instrument-Big Science. Where does instrumental realism fit in here?

We have already seen two of the issues which will be of importance as questions which arise from experiment come forth: Virtually all our instrumental realists have noted that in the "carpentering" of an experiment and in the development of instrumentation, there are junctures at which there may be high ambiguity about whether what is observed is "real" or an artifact of the instrument/experiment itself. In short, is the entity in question itself a "construct" of the instrumental complex?

Our realists usually agree that whether through the refinement of the instrumentation or through a multiple set of variations via several instruments, there is a point at which the entity may be pronounced "real." Hacking sets the tone in *Representing and Intervening.* He argues that there is some point at which the intervening "observer"—now become a kind of *technological manipulator*—gets to the things themselves. In the context of the realism/anti-realism debate, Hacking noted:

> For my part I never thought twice about scientific realism until a friend told me about an ongoing experiment to detect the existence of fractional electric charges. These are called quarks. Now it is not the quarks that made me a realist, but rather electrons. Allow me to tell the story [Hacking follows with the story of the manipulation of the experiment which allows for altering charges on a target]. . . . Now how does one alter the charges on the niobium ball? "Well at that stage," said my friend, "we spray it with positrons to increase the charge or with electrons to decrease the charge." From that day forth I've been a scientific realist. *So far as I'm concerned, if you can spray them then they are real.*[7]

Here the referential realism has been changed from simple observation to carpentered intervention on a virtual tactile-manipulative model. The scientific objects are being manipulated, albeit indirectly through the instrumental set. Here we have the clue to a second level of the embodiment of science in technologies—the scientific phenomenon is not only accessible through manipulation but occurs only in the context of a carpentered instrumental context.

Heelan, too, returns to the fore here, and what appeared first as his narrowness—scientific entities are those entities which appear *only in carpentered environments, technologically constructed*—becomes a virtue. Such carpentering is the normative mark of a Big Instrument science. Both Heelan and Hacking, while remaining instrumental *realists* retaining an embodied referential realism, are clearly at the juncture of the transformation from the "observational" to the "laboratory" shift of model for the center of Big scientific praxis.

I shall return to our instrumental realists shortly; however, regardless of how aware of the role of observation within the context of Big Instrument science Heelan and Hacking may have been—and they now are working on follow-up books addressing these problems—it remains the case that two new thinkers have addressed and foregrounded the issues before the completion of the next stage of instrumental realism. I refer to Peter Galison and Bruno Latour, the former addressing the issue from a historical perspective, while the latter took a more sociological view.

The 1987 publication of Peter Galison's *How Experiments End* was another high-water mark in the now established shift in the new history and philosophy of science. Galison combines the multiple disciplinary credentials of philosophy, history, and physics, and while *How Experiments End* is narrow—it deals only with a certain history of physics experiments—it is highly instructive in this context.

Galison's opening echoes all the shifts of emphasis previously noted: (a) He recognizes and counters the old overemphasis upon theory disembodied from experiment, "Despite the slogan that science advances through experiments, virtually the entire literature of the history of science concerns theory . . . the histories . . . highlight the evolution of concepts, not laboratory practice";[8] (b) He criticizes the old philosophy of science's notion that the operations of science are those of logical procedures, "There is no *strictly logical termination* point inherent in the experimental practice. . . . Experimentalists' demonstrations of the reality—or artificiality—of an effect or particle will never have the closed form of a deductive argument";[9] and (c) He implicitly argues for a praxis orientation—although I do not class Galison with the praxis-*perception* model of interpretation directly—which is situated in a three-part relation, "We . . . use this threefold interaction of experimental practices, theory, and instruments to explore features of the huge accelerator based particle experiments that symbolize our . . . epoch."[10]

What Galison does is trace out a fairly long history of physical experiments and concentrate upon the emergence of Big Instrument physics at present. In the process he marks significant changes in both the practice and construction of the experiment. While it is a masterful book with many implications for the parallel rise of instrumental realism, I shall concentrate here only upon several highly suggestive aspects of the history-praxis of the experiment as it bears upon the interface of science and its technologies.

An overview of *How Experiments End* produces the first parallelism and confirmation of the higher-level embodiment of science in its technologies. Galison traces three stages of physics experiments, beginning with the atomic theory development of the nineteenth century. He notes that at that stage, physics could investigate only macro-phenomena with moderately simple equipment, often yielding only inferential (hermeneutic) results.

> At the time (19th century) experiments on microphysics relied on inferences from the behavior of macroscopic quantities of atoms . . . Typical

> of these macroscopic experiments was James Prescott Joule's work on friction in the 1840's to demonstrate that heat was just motion at the scale of atoms. Other macroscopic experiments such as spectroscopic investigations helped in the construction of atomic models.[11]

Here we have experiments constructed with relatively simple machinery which, due to the macroscopic referent, is not capable of either fine or massive intervention or manipulation. Were one to formulate a continuum from "observation" to "manipulation," these experiments would remain nearer the observation end—although the "readings" obtained belong primarily to what I have called hermeneutic relations.

The second stage in Galison's history escalates technological sophistication with a subsequent reduction in the size of phenomena which can be investigated:

> The first epoch of macroscopic experiments was gradually replaced in the 1910's and 1920's by a second epoch in which atoms could be examined one by one. Building on the late nineteenth-century discovery of radioactivity, physicists designed instruments that were sensitive not to the macroscopic effects of heat, light, mechanics, and electricity but to the passage of individual rays and particles.[12]

Here the refinement of the instruments, with the subsequent ability to deal with more microscopic levels of phenomena, also comes closer to micro-manipulations. By the 1930s atomic theory had undergone the theoretical revolutions of relativity theory and quantum mechanics—both of which arose in conjunction with the earlier epochs of observation and experiment at the macroscopic stages noted. But with the ability to experiment at the level of the atom, instead of its macroscopic effects, the door was opened to the third epoch of *particle* physics.

But to get at particles, new instruments were needed—and these had to be more sophisticated and able to manipulate matter on a smaller scale than any previous technologies allowed. Enter yet another generation of instrumentation, ultimately that of the macro-instruments, CERN and Fermilab accelerators on the largest scale, and the competing, but equally sophisticated and complex Gargamelle heavy liquid bubble-chamber instruments. What Galison's fascinating and careful history shows is a three-fold movement in the development of the contemporary physics experiment:

1) There is a clear movement from "observation" toward "manipulation," as previously noted by Hacking and Rouse. The older model of science as observation, toward the high-technology laboratory, from what might be called an "astronomy" configuration to an "engineering" configuration, moves with the evolution of the experiment through the three epochs Galison traces.

2) But this movement also *confirms* the previous analysis of technological trajectories I earlier traced in *Technics and Praxis,* but with more significant detail. One could say that the simple, "micro-technologies" of early physics were not yet capable of manipulation of natural phenomena in any powerful or significant way. The larger the macrophenomenon and the smaller the instrument, the more limited one is to the mere "observational" model. By contrast, once the instrumentation becomes sophisticated and complex enough to reach down into the more microphenomenal levels (often also related to instrumental size), the more manipulative the experiment may and does become—Big Instrument manipulates microphenomenon.

3) This movement also follows what I have called a *technological trajectory* and exemplified a subtle way in which science may well *follow* the implicit suggestibility of technological possibilities within science's own technologies:

> It was noted that instrumental mediation transforms the shape and distance of the world. The desevering or bringing into presence of that which was previously either unnoticed or undetected is done in such a way that micro-macro features of the world are made focal. The instrument, metaphorically, concentrates upon the micro-macro features of the world. This structural component of instrumental transformation was, of course, noted as latent in the previous characterization. But here this feature taken as a telic possibility of instrumentally embodied investigations becomes more obvious as an actual feature of contemporary science.
>
> That contemporary scientific investigations are highly committed to the examination of extreme micro-features is abundantly apparent. Particle theory in physics, DNA and genetic investigations in biology, and the construction of elements from micro-phenomena in chemistry are all familiar. What is not noted so strongly is that these areas of investigation are frequently regarded as the highest prestige areas of frontier research in the sciences. Such investigations are, of course, not possible without increasingly sophisticated and—perhaps incidentally-*large* instruments. My suggestion is that not only is the condition of the possibility of such investigations related to technics, but that the latent telos of such technics provides in part the ground for such investigations.[13]

Thus, Galison's analysis can be seen to return to and confirm one of the important insights arising from the interface of the philosophy of science with the philosophy of technology.

From the perspective of gaining scientific knowledge, one might say that the evolution of larger, more sophisticated technologies which, as Ackermann contends, allow actual progress for such knowledge, is a clear gain. But it is not an unambiguous gain. Indeed, if it is a gain, it now becomes very enigmatic.

Galison's title is itself suggestive of this ambiguity—how do experiments end? What becomes clear in Galison's history of physics experiments is that there is no obvious or unambiguous *end* to experiment. Instead, there is an implicit threat to the vary "aha!" phenomenon which often signals the perceptual gestalt which our instrumental realists value. Is realism, even *instrumental* realism, threatened?

There is an overarching fourth movement in the Galison three-epoch history of physics experiments. In this movement there is a diminishment of certainty or even convincing satisfaction with the experimental result as the experimental context itself becomes more complex and manipulative with respect to the microphenomena.

At stage two, the stage at which atoms may be investigated, Galison argues that the end of experiment becomes a series of ends, with diminished "moments of discovery":

> Instead of looking for a "moment of discovery," we should envision the ending of the muon experiments as a progressively refined articulation of a set of phenomena. In a sense the experiment had to end several times. At each stage of the process, a new characteristic could be ascribed to the cosmic rays; they discharged electroscopes; the discharge rate varied in a certain fashion with depth in matter; the shower particles were more easily absorbed than the single particles. . . . [14]

This conclusion, in which the as yet unclear entity is restricted to its instrumental behavior, exactly parallels the insight of Latour, to whom we shall turn shortly. "The new object, at the time of its inception, is still undefined. More exactly, it is defined by what it does in the laboratory trials, *nothing more, nothing less.*"[15] And although Latour's context is different from Galison's, here again is an independently arrived at conclusion regarding current experimental practice among our praxis-oriented thinkers.

Then, what remains at the end of epoch three is even less clear. After "neutral currents" come into and pass out of being the ambiguous "end" of an experiment becomes itself a dismantling of an experiment, with results cast negatively, as "one does not see how these effects can be made to go away." The physics community had been conducting a series of experiments with different instrumental complexes, but results remained weak.

What emerged was a concatenation of (a) groups seeking to bolster support for papers to be published, (b) results from multiple instruments, and (c) distinct complex results being brought into agreement. Concerning the first point Galison notes:

> By the tone and context of these memorandums [sic] it is apparent that these considerations served not to *persuade* members of the collaboration—by late February everyone believed that neutral currents would not "go

away." These final arguments were intended rather to bolster the public version of the paper by minimizing dependence on the calculated quantities of the computer simulation.[16]

Refinements led, finally, to some stabilization of the data. "The data were stable. Three different muon detectors gave concordant results . . . ," and, "Here was a final reconciliation of the two experiments, 'even though the corrections differ greatly in magnitude and technique. . . . ' "[17] When published results at last appeared—dressed by the above processes and confirmed by later repeated experiments—"the existence of the effect seemed assured."[18]

Galison summarizes the complexity of the result:

> So it was that the early neutral-current experiments drew to a close. Two diverse collaborations, each with its own internal dynamic of expectations and expertise, had found their way to a conclusion, integrating conflicting approaches, understanding machinery, learning to reclassify phenomena. On both sides of the ocean, groups of physicists had forced themselves to come to terms with a multitude of analyses, objections, and proposals raised from among their own ranks. Over and over, both groups pushed their experiments toward a more direct contact with measured quantities, and struggled with their results until they exhibited a measure of stability—all under enormous pressure . . . the Europeans finally mastered what amounted to two experiments . . . so too the Americans produced more than one demonstration.[19]

One could say that what emerged was a science pressured by consensus—but not without its immersion in the technological context of instrumentation, the machinery which *produced* the data for the consensus. The complex of high-technology experiment sets the stage for the sociology of science.

As indicated earlier, in the late 70s, but particularly in the 80s, the sociology of science began to gain strength. David Bloor's *Knowledge and Social Imagery* (1976) was an early book that applied what had long been the notion of the "social construction of reality" among social scientists to natural science. Later, more specific examinations of laboratory science were made by Karin Knorr-Cetina in *The Manufacture of Knowledge* (1981), and by A. Pickering in *Constructing Quarks* (1984).

But the thinker who most radically poses the "social construction" thesis is Bruno Latour in *Science in Action* (1987). I tend to agree with his *Times Literary Supplement* reviewer that *Science in Action* may well be the sociology of science equivalent to Kuhn's *Structure of Scientific Revolutions.* Yet, there might seem to be initial problems with relating a view which places "reality questions" concerning science in the background.

In its most radical form, social constructionism brackets any questions of reality in the form in which philosophers of science might be interested. Even the realism of instrumental realism would have to take a background role. Yet, there are deep aspects of Latour's approach which make it very proximate to these investigations: (a) His is clearly a *praxis* analysis, although in a form closer to the later Foucault's type of analysis, a discourse and power praxis. (b) He takes a perspective upon technologically embodied science—which he calls *technoscience*—that allows the necessary interrelation of science and technology to be foregrounded. And, in what I shall term a "sociological epochē," (c) He inverts the usual perspective upon technoscience in such a way that many of the themes I have followed concerning both praxis and perception can be seen in a new light.

Latour's method strikes me as having a deep analogue to procedures we have noted previously arising from phenomenology, although without their direct perceptualist emphasis. He focuses upon the practices of science: "We must stick . . . carefully . . . to our method of following only scientists' practice, deaf to every other opinion, to tradition, to philosophers, and even to what scientists say about what they do."[20] And, what I am calling his "sociological epochē" inverts the usual order about what text-book science might be thought to be.

A "natural attitude" text-book portrayal of science might claim that Nature "poses" questions (and ultimately answers them) to scientific inquiry. Then, through the application of experimental method—instantiated in the *laboratory* experiment—its truths are discovered and recorded in scientific texts, journals and other publications report the results of the inquiry.

Latour inverts this progression by beginning with the literature of science, moving backward and behind it to the laboratory, and eventually arriving at "Nature." One might call this an archeology—after Foucault—but it is an archeology with a difference. This is not simply an inversion of order, it is an inversion of how the entire praxis if viewed. The new perspective which Latour establishes revolves around a series of Janus-faces which are illustrations of his deliberate inversions, all of which distinguish science *in action.* They strongly distinguish between science prior to the settling of a dispute, and science sedimented, after the settling of a dispute. Here one can easily recall both Kuhn's revolutionary science and, more recently, Galison's analysis of how experiments play their role in the settling of disputes—indeed, there are very many good parallels between Galison and Latour, in their rich empirical histories, and even between them and the outcome of experimental histories.[21]

However, Latour's Janus-faces *privilege* the moments of the unsettled parts of technoscience and, in effect, read settled science from that privileged perspective. Because of this, Nature is always the last to arrive. From the point of view of developing praxis, " . . . it is Nature who always arrives late, too late to explain the rhetoric of scientific texts and the building of laboratories."[22] From the point of view of settled or sedimented science, Nature is the cause of that which allows controversies to be settled—the whole rhetoric of science retrospectively and traditionally makes that claim—but from the perspective of the movement to settlement, Nature will be the consequence of the settlement. "Nature is never used as the final arbiter since no one knows what she says. But *once the controversy is settled,* Nature is the ultimate referee."[23]

Latour makes any initial appeal to Nature a consequence rather than a cause: "Since the settlement of a controversy is *the cause* of Nature's representation not the consequence, *we can never use the outcome—Nature—to explain how and why a controversy has been settled.*"[24] Such an inversion serves simultaneously to reveal a new perspective—a function again very similar to a phenomenological epochē—and to highlight praxis. Nature is sardonically characterized by Latour:

> Nature, in scientists' hands, is a constitutional monarch, much like Queen Elizabeth the Second. From the throne she reads with the same tone, majesty and conviction a speech written by Conservative or Labour prime ministers depending on the election outcome. Indeed, she *adds* something to the dispute, but only after the dispute is ended; as long as the election is going on she does nothing but wait.[25]

The purpose of such an inversion is to foreground praxis, the process of "the fabrication of scientific facts and technical artifacts."[26] Note that Latour always groups and parallels scientific *facts* and technical *artifacts.* Part of his perspective is the functional combining of science with its technologies into *technoscience.* From this perspective, such a view has the advantage from the outset of seeing science as embodied. Latour's purpose, of course, is different from mine—it parallels Galison's interest in the corporate and development of scientific controversy, the *socially constructive* aspects of technoscience. "To sum up, the construction of facts and machines is a *collective* process."[27]

Yet while technoscience is clearly an embodied science, its material products (and established "facts") are backgrounded. For just as Nature is what is "added" as consequence, so also are these products the after-effects of science in the making. Latour introduces another inversion: "The fate of facts and machines is in later users' hands; their qualities are thus a consequence, not a cause, of a collective action."[28]

Such counter-intuitive inversions have the distinct advantage of making the collective and the developmental aspects of technoscience stand out. Looked at in this way, technoscience begins to appear as a very complicated socius (or 'machine') for creating and resolving controversies, with the focus upon *technoscience in action.*

The first level of this praxis examined by Latour is that of the literature of technoscience. And although I shall not dwell on what turns out to be a most fascinating analysis and instead concentrate upon what he has to say about instruments and the laboratory, it is worth indicating some of the parallels with Galison.

Both Galison and Latour are aware of the *rhetorical and political* dimensions to technoscience publishing. Indeed, the literature of technoscience is one place in which the *contests of strength* which Latour highlights come to the fore. However, Latour's perspective has a higher contrast precisely because of his deliberately practiced inversions which include:

1) "When controversies flare up the literature becomes technical."[29] This is an inversion of the usual common-sense approach to science which holds that tight, technical language is designed to *lessen* controversy (it does serve to mystify outsiders and exclude them from the tribal languages).

2) Within the technical approach, Latour shows that technoscience modifies many very ancient and traditional rhetorical devices, including appealing to authority, gathering multiple witnesses, referring to former texts, and tightening a network. Citation practice is a crucial part of this whole process.[30]

3) Stylistically and in other ways, technoscience's rhetoric is built in such a way as to withstand attack (it belongs to the process of controversy building and resolution). Only when the initial garnering of support succeeds and the response is genuinely cited and accepted can the result be established or considered a "fact." Thus Latour's rule "asks us not to look for the intrinsic qualities of any given statement but to look instead for all the transformations it undergoes later in other hands. This rule is the consequence of what I called our first principle: the fate of facts and machines is in the hands of later users."[31]

4) But most interesting is Latour's characterization of the peculiarity of technoscience writing in the context of conflict production and resolution:

> What I will call *fact-writing* in opposition to fiction-writing limits the number of possible readings to three: giving up, going along, working through. *Giving up* is the most usual one. People give up and do *not* read the text . . . [sometimes] because they are pushed out of the controversy altogether . . . [Latour estimates this option at 90 percent]. *Going along* is the rare reaction, but it is the normal outcome of scientific rhetoric . . . [Or, the rarest, *working through* is] such a rare and costly one that it is almost negligible as far as the numbers are concerned; *re-enacting* everything that the authors went through.[32]

All of this, Latour claims, is the peculiarity of the scientific text which "[chases] its readers away whether or not it is successful. Made for attack and defence, it is no more a place of a leisurely stay than a bastion or a bunker. This makes it quite different from the reading of the Bible, Stendhal or the poems of T. S. Eliot."[33]

But, for the rare challenger, the place for the challenge is not literature (which comes later), but the *laboratory.* Here the stakes get higher, "If you dispute further and reach the frontier where facts are made, instruments become visible and with them the cost of continuing the discussion rises. It appears that *arguing is costly.*"[34] One might also note that given the relative cost of science publications—sometimes running to nearly $2,000 per year for some periodicals, compared to the usual expense of less than $100 per year for philosophy journals—the cost is *already high.* But with the need of a laboratory, the casual reader or the low investment are excluded, thus the difficulty of external challenge is steeply increased.

Technoscience's literary referent is the *laboratory,* with its enactment of the experiment. If challenges are to be made, they will be carried out in, or through, the laboratory. Laboratories are the higher stake and the material embodiment of the tests of strength which Latour claims characterizes technoscience's praxis. Ultimately, the cost and necessity of a full challenge in a test of strength may have to lead to a *counter-laboratory,* complete with its own armies of researchers and contestants.[35]

It is with the laboratory and its instruments that *Science in Action* makes strong contact with the interests of our instrumental realists. And in similar refreshing originality, Latour re-characterizes the experimental context. I shall here concentrate upon only those elements which bear most directly upon the themes of instrumental realism:

First, the laboratory is simply the (high-cost) place where scientists work. But in an interesting twist, Latour describes the purpose of the laboratory as a place where *inscriptions* are made. The laboratory is "like" a kind of writing, in Jacques Derrida's sense.[36]

> When we doubt a scientific text we do not go from the world of literature to Nature as it is. Nature is not directly beneath the scientific article; it is there *indirectly* at best. Going from the paper to the laboratory is going from an array of rhetorical resources to a set of new resources devised in such a way as to provide the literature with its most powerful tool: the visual display.[37]

Then, with high originality, Latour defines the instrument as an *inscription-making device for such a visual display:*

> This move through the looking glass of the paper allows me to define an *instrument,* a definition which will give us our bearings when entering

> any laboratory. I will call an instrument (or *inscription device*) any set-up, no matter what its size, nature, and cost, that provides a visual display of any sort in a scientific text.[38]

Here we are smack-dab right back to the praxis-perception model explicit or implicit in instrumental realism, although we have arrived there through a very different route. We are also at the point introduced earlier in *Technics and Praxis* and elaborated much more fully in *Technology and the Lifeworld.* Instruments which produce "readings"—those which exemplify *hermeneutic* relations—are indeed modelled upon reading practices. We have seen this move earlier in Ackermann's and Heelan's work, as well as in my own.

I shall not argue that Latour belongs entirely to the "school" I have been dealing with, since he backgrounds some of the issues our group foregrounds. But I can indicate how he fits nicely into one whole set of these concerns:

1) If the instrument is a device for providing an inscription in a visual display, it is clearly what Heelan calls a "readable technology." For Latour, the thin line between laboratory and text occurs at the juncture of the instrumental read-out and the textual graph. The imaginary context within the lab has the PI explaining to the visitor how the instrument provides the material for the published graph:

> We thus realize where this figure comes from. It has been *extracted* from the instruments in the room, *cleaned, redrawn, and displayed. . . .* We also realize, however, that the images that were the last layer in the text, are the *end result* of a long process in the laboratory that we are now starting to observe. . . . For a time we focus on the stylus pulsating regularly, inking the paper, scribbling cryptic notes.[39]

2) But the reading of instrumentally produced inscriptions—the Derridean "writing" of the laboratory—requires a long praxis background of training and production. "In order to argue, we would now need the manual skills required to handle the scalpels, peel away the guinea pig ileum, interpret the decreasing peaks, and so on."[40] The hermeneutics of technoscience entails this long training and bringing to skill of its "reading techniques."

Such skilled reading is very different for the outsider, compared to the insider. Latour makes this contrast vivid:

> We are no longer asked to believe the text that we read in *Nature;* we are now asked to believe *our own eyes. . . .* The object we look at in the text and the one we are now contemplating are identical except for one thing. The graph of sentence (1), which was the most concrete and visual element of the text, is now in (2) the most abstract and textual element in a bewildering array of equipment. Do we see more or less than before? On

> the one hand we can see more, since we are looking at not only the graph but also the physiograph, and the electronic hardware . . . [etc.]. We can see more, since we have before our eyes not only the image but what the image is made of. On the other hand we see *less* because now each of the elements that makes up the final graph could be modified so as to produce a different visual outcome.[41]

3) This is "reading" which entails perception, but of a very special kind previously recognized by our instrumental realists. But it is now dominantly a kind of reading which is often far from isomorphic, *embodiment relations,* as I have called them. Many of the visual displays are the *hermeneutic relation* sort:

> With a minimum of training we seek peaks; we gather there is a base line, and we see a depression in relation to one coordinate that we understand to indicate the time. . . . When we are confronted with the instrument, we are attending an "audio-visual" spectacle. There is a *visual* set of inscriptions produced by the instrument and a *verbal* commentary by the scientist.[42]

Latour notes that for beginners this is always impressive because the guided *demonstration* appears rich and full, with the appearance being that of the scientist saying only what the inscription requires. Yet, for the trained scientific hermeneut, this can also be the opening for challenge, for a species of hermeneutic criticism, since, without the commentary, "the inscriptions say considerably less!"[43]

4) But the form which a critical hermeneutic must take in the laboratory is also different from that of the library. It calls for the actual physical and bodily praxes which characterize technoscience. In the laboratory the hermeneut who will challenge a "reading" opens a "trial of strength."

Latour cites an example from epoch two of Galison (the epoch of manipulating atoms, that of the discovery of X-rays, etc.). Blondlot, a French physicist, claimed to have discovered a new ray which he called an N-ray, and constructed instruments needed to show effects upon plates. But Robert W. Wood of the U.S.A. doubted the result and issued a challenge which entailed actually manipulating the instrument. "At one point he even surreptitiously removed the aluminum prism which was generating the N-rays. To his surprise, Blondlot on the other side of the dimly lit room kept obtaining the same result on his screen even though what was deemed the most crucial element had been removed."[44] The end result of the challenge, the demythologizing of the reading, was a classic Kuhnian shift, "After Wood's action . . . no one "saw" N-rays any more but only smudges on photographic plates when Blondlot presented his N-rays. . . . The new fact had been turned into an artifact."[45] We have seen precisely this ambiguity before. It frequently occurs through readable technologies. Like texts, instrumental displays call for both critical and trained reading praxes.

In the above context, Latour's deconstruction of scientific text/ instruments remains consonant with that of the instrumental realists in spite of his quite different focus upon social construction. But the stakes are once again raised when both the cost and challenge within Big Instrument/Big Science rise to the level of the building of counter-laboratories. (Of particular interest for *Science in Action* is the historical fact that the controversy over "cold fusion" arose after Latour's analysis. Yet his analysis about all levels—literary rhetoric, demonstration, and even the building of counter-laboratories—serves as an excellent, even *predictive* diagnosis of what has happened in the "cold fusion" debates to date!) "[With] all else being equal, there is no other way open to the dissenters than to *build another laboratory.*"[46]

Latour sees this escalation as precisely parallel to the contestative model noted in the literary dimension:

> The price of dissent increases dramatically and the number of people able to continue decreases accordingly. This price is entirely determined by the authors whose claims one wishes to dispute. The dissenters cannot do less than the authors. They have to gather more forces in order to untie what attached the spokesmen and their claims. This is why all laboratories are *counter-laboratories* just as all technical articles are counter-articles.[47]

Latour's focus remains the socius and the construction of scientific fact and instruments. Yet the crucial issues remain those which are consonant with the concerns of the instrumental realists. For what happens for the successful counter-laboratory is often the construction of a more discriminating "visual display" machine. Latour illustrates this himself. In the Schally controversy over GHRH, his competitor Guillemin invents a new process:

> The new assay is much *more* complicated than Schally's older ones . . . it gives inscriptions at the end that may be said to be more clear-cut, that is they literally cut shapes out of the background. In other words, even without understanding a word of the issue, the perceptive judgment to be made on one is easier than on the other.[48]

What the new counter-laboratory does is escalate the complexity and the cost of the display. Its method of fact *and machine* construction and fabrication is raised. Latour's point is that this escalation more and more excludes challengers while simultaneously guaranteeing apparent success.

What kind of success? The test must be to determine what kind of "reality" is produced. It is at this juncture that not only our instrumental realists, but Galison also, meet with Latour. We have already seen that Nature is backgrounded, that both facts and instruments are fabricated; but what is the ultimate "reality" that is constructed?

The answer is, first, an object, then, a thing which can stand forth independent of the laboratory and the process which fabricated it. From the point of view of both the scientist and the usual philosopher of science, Latour's *object* would be a new "scientific object," such as a neutral current or a quark. But, following Latour's emphasis upon the process of settling contexts of strength, let us take a different perspective:

> We have reached a point which is . . . most delicate . . . because, by following dissenting scientists, we have access to their most decisive argument, to their ultimate source of strength. Behind the texts, they have mobilized inscriptions, and sometimes huge and costly instruments to obtain these inscriptions. But something else resists the trials of strength behind the instruments, something that I will call provisionally a *new object.*[49]

At first, the object is defined behaviorally: "It is defined by what it does in the laboratory trials, *nothing more, nothing less;* its tendency to decrease the release of growth hormone . . . [etc]."[50] Inside the laboratory, this new object, defined by its behavior, is *"a list of written answers to trials."*[51]

Here we are at the stage near the end of an experiment, noted previously by Galison. The data are stabilizing and the object is close to becoming a *thing.* "The "things" behind the scientific texts are thus similar to the heroes of the stories . . . and are all defined by their *performances.*[52] But with repeatability in the experiment and the gradual ability to further manipulate the object, one notes that "each performance presupposes a *competence* which retrospectively explains why the hero withstood all the ordeals."[53] We are near a *product.*

The final act, for our purposes, is the transformation of this new object into a *thing* able to exist independently of the laboratory. The laboratory is a *scientific factory,* its *technological production facility.*

> What makes a laboratory difficult to understand is not what is presently going on in it, but what *has been* going on in it and in other labs. Especially difficult to grasp is the way in which new objects are immediately transformed into something else. As long as somatostatin, polonium, transfinite numbers . . . are shaped by the list of trials . . . it is easy to relate to them: tell me what you go though and I will tell you what you are. This situation, however, does not last. New objects become *things:* "somatostatin," "polonium," . . . "double helix," or "*Eagle* computers," things isolated from the laboratory conditions that shaped them, things with a name that now seems independent from the trials in which they proved their mettle.[54]

Here the payoff of Latour's isomorphism of fabricated "facts" and "machines" begins to emerge: the technoscientific product which is constructed is more like a technological object than a "natural" object.

> This process of transformation [of the laboratory-produced "object" into an independent "thing"] is a very common one and occurs constantly both for laypeople and for the scientist. All biologists now take "protein" for an object; they do not remember the time . . . when protein was a whitish stuff that was separated by a new ultracentrifuge . . . [when] protein was nothing but the action of differentiating cell contents by a centrifuge.[55]

Latour's social construction demystifies the end result: "This process is not mysterious or special to science. It is the same with can openers we routinely use in our kitchen."[56] Of course, the access, the instrumentarium, and the skills required to "read" the displays, are all very different from the level of use of the can opener.

What comes into view, however, is an interesting crossing of what could be called the convergence of Nature with Culture through laboratory science. The object—produced in the laboratory and reified into an independent thing—was fabricated, but fabricated in such a way that it takes on the status of "reality." "Laboratories are now powerful enough to define *reality*."[57] Latour, again demystifying, argues:

> To make sure that our travel through technoscience is not stifled by complicated definitions of reality, we need a simple and sturdy one able to withstand the journey: reality as the latin word *res* indicates, is what *resists*. What does it resist? *Trials of strength.* If, in a given situation, no dissenter is able to modify the shape of a new object, then that's it, it *is* reality, at least for as long as the trials of strength are not modified.[58]

Here we reach a point at which Latour and Galison—at least functionally—converge. The "effects cannot be made to go away." Reality is what resists, at least until more powerful instruments and laboratory complexes can overthrow them.

Yet beyond Galison, Latour introduces a suggestion which extends the previous continuum from observation to manipulation—*now to the production of entities.* For if pre-modern science was limited to observation, and even early Modern science operated on what I have called an observatory constellation, and if Modern science became increasingly a manipulative and intervening science, then Postmodern science, *technoscience,* becomes a *productive* science in precisely the laboratory/factory sense.

This is no way removes it from the close and necessary embodiment of science in its technologies. To the contrary, it blends the roles of science and technology such that it is virtually impossible to differentiate between these functions. And while physics remains the most costly and complex science—thus remaining at the top position in the trials of strength from an economic and social perspective—there are at least indications that the new technosciences of *biotechnology* may even better exemplify emergent Postmodern science.

The constructed, the manufactured entities of micro-manipulated entities are precisely the organic/inorganic junctures, and the cloned and gene-manipulated entities being produced are more clearly the products of the Nature/Culture ambiguities now almost paradigmatic of Postmodern science.

As the biologies (genetic, biochemical, technobiological) flex for the new trials of strength, they too begin to build Big Laboratories and counter-laboratories, following the previous Big Science leads of physics and chemistry; but they are perhaps better focused upon the manipulative/productive models emergent here.

At least, this is the suggestion implicit in the Galison/Latour modes of praxis interpretation. By way of postscript, the perceptualist and materially contexted insights of our instrumental realists continue to bring forth and remind us of the *technological* reality of Postmodern technoscience. Produced entities comes into being in the "real" world. And "manufactured" entities can also be "painted." But they can't so easily be differentiated from pure "Nature"; they belong, at best, to the juncture of Nature and Culture. We are a very long way from the Old philosophy of science.

Epilogue: Philosophy of Technology beyond Philosophy of Science

The focus upon the interface between philosophy of science and philosophy of technology through the technological embodiment of science in instruments is an important but *narrow* focus. It is important because as argued by our instrumental realists, it is really science's technologies which have been much of the basis of uncovering the new. This is particularly the case in the now large dimensions of the macro- and microlevels of reality beyond the reaches of unaided perception at the very base of scientific progress. This new or re-balanced focus as opposed to the preoccupations of an older philosophy of science is intrinsically important. And when this leads, as it has here, to the ambiguity of the scientific object which is "real" but produced, then the incarnation of science as *technoscience* is even more marked.

Yet the focus is also narrow. This is because, even with instrumentation, there are related and secondary phenomena which also need to be noted. And as the location of a region of overlap between philosophy of science and philosophy of technology, this interface as developed is only suggestive.

At the beginning, I remained within the umbra of science's instrumentation and its effects. The dominant view of science held by most philosophers would hold that the primary trajectories of a science-technology relation are those which can be circumscribed by the overall idea of a *science-driven technology.* This occurs at the simplest level in the notion of "pure" science eventually producing some "applied" effect. I began this primer by questioning that set of priorities. By now it should be clear that there is another direction of effect; there is also a *technology-driven science.*

At the highest altitude, such a perspective was suggested most radically by Heidegger, who holds that what we take to be science—even in its most theoretical heart—is an effect of a technological way of taking things, of "revealing a World." But at a lower and much more concrete level we also have noted how parts of our world are instrumentally and technologically revealed and even produced.

Instruments, our instrumental realists have pointed out, are the essential means for that world-revelation. Yet here too the focus has been perhaps too narrow. At the beginning, even if recognizing more complexity in principle, the choices of illustrations concerning instruments may have been overly cautious and simple. Most of the examples in the first books are a little like Heidegger's equally favored selection of simple tools in the workshop, with associated fears of larger technologies such as dams on the Rhine or atomic bombs. While *fears* about larger and more complex instrumentation do not trouble our authors, there remains something of the smaller and more manageable to the examples selected by all of our instrumental realists. Telescopes and microscopes—even if updated in high-tech ways to include electron and sonic microscopy and spectral and light-enhanced telescopy—remain central. And with the hermeneutic variety of indirectly "read" instruments, we remained with thermometers or display panels.

Not that more could be done with even these examples: Use and design were dealt with differently by different authors. And one secondary area of what I call *technology-driven science* also was located. Some, and increasingly many scientific phenomena are clearly *technologically carpentered* phenomena. Heelan makes this sort of carpentry central, but Hacking and Ackermann also recognize at least one class of scientific phenomena which would not be known or would not exist without technologies. And most recognize the historian's adage about "science owing more to the steam engine than the steam engine to science." One of the carpentered examples is clearly that of the laws of thermodynamics arising in relation to steam engine performance rather than nature observation. The same technology-driven modelling is taken to be excessive in Dreyfus's critique of artificial intelligence. He clearly doubts that the laws of psychodynamics will arise in conjunction with computer modelling. But beyond sensory enhancement lies a whole realm of technology-artifact-produced science. This, too, is part of the interface between philosophy of science and philosophy of technology in the shadows of instrumentation within the large laboratory.

These phenomena lie within the direct examination of instrumentation, and they have been noted in various ways by each of our authors. But there is also a more subtle, secondary effect which ought to be pointed up as belonging to technology-driven science. It is what I call the *inclination of a trajectory.* Such inclinations are related to the capacities opened up by instruments, capacities of a technological possibility leading to the productive capacities of experimental science.

At the highest and most general altitude, if it is true that research programs in the sciences are more and more concerned with the macro- and microlevels, is not this itself an indicator of following a technologically possible trajectory? Does the array of instrumentation "suggest" just such a direction? Even in the case of our simpler instruments, just such a trajectory may be detected in the early stages. Magnification "suggests" more magnification; resolution more resolution, until eventually, we reach not only the historic refinements of microscopes and telescopes, but their contemporary variants which also present whole-image results isomorphic with ordinary vision—NMR, songrams, etc. This is following a technological trajectory with its fascination. But it is also a subtle indirect and secondary effect of a technology-driven science.

In the just-cited examples, we remain within the narrowly focused domain of science's instrumentation—but science is more than technologically embodied. It is also institutionally *technologically embedded.* That was clearly recognized both by the new generations of discourse and power praxis philosophers of science such as Rouse, Gutting, et al., and by the sociology of science thinkers. Here we focused upon such praxes in the laboratory, the experiment. To Galison and Latour I could have added others who have not yet produced books. Indeed, in this series Robert Crease now joins the writers in experiment, with his *The Nature of Scientific Experiment.*[1] Today's Big Science is so closely tied to Big Technology that one can meaningfully speak of a single, complex phenomenon whch is both a scientific technology and a technological science: *technoscience.*

Not that our instrumental realists have overlooked this complexity within Big Science. Ackermann had already accounted for some of the social construction in science and its embeddedness in the political matrices of our times. Heelan and Hacking are today engaged in book projects focusing precisely upon the larger socius of experiment and corporate science.[2] And my recent *Technology and the Lifeworld* places science in the context of multiple world cultures.[3]

There is even a sense in which the embeddedness of science in corporate modes is as old as its beginnings in the Renaissance, although without the levels of complexity and megascale technologies as in the last five decades. One finds our Galileos and da Vincis courting, in anticipatory fashion, precisely the equivalents of today's "military-industrial complex," about which Eisenhower warned us. At its birth, one could say that science foresaw itself as power- and money-oriented. Big Science is corporate-structured and concretely overlaps in both style and organization other corporate structures of clearly less "theoretical" bent. These factors which today disturb social consciousness are part of the context for the internal modes of contestation and exclusion which the hermeneutic philosophers of science and the sociologists of science detect in the high-ante stakes within laboratory science.

This fact points to an entirely different interface area between philosophy of science and philosophy of technology but one which does not so neatly overlap, as in the examples explored more deeply here within the confines of epistemology and ontology. The issues surrounding Big Science-Big Technology are unavoidably linked to social-political and ethical philosophies in ways which go beyond most extant philosophy of science.

If the dominant strands of philosophy of science have heretofore been insensitive to or forgetful of the science/technology interface in instrumentation, these same strands have been equally negligent with respect to the social structures and operations of science in its now dominant corporate form (excluding some strands of Popperian, Marxian, Critical Theory, and Feminist philosophers also recognizable as important minoritarians). Within the philosophy of science, only recently and primarily from the discourse-praxis and hermeneutic philosophers of science, have these issues been seen to be deeply intrinsic to the institutions of science itself. But ethics, social-political philosophy, and concerns for social effects *frequently have been central to much philosophy of technology.*

In short, much philosophy of science has concentrated upon the process of discovery, and that, within fairly narrow boundaries. It might be said, in contrast, that in its more dominant concerns, much philosophy of technology has concentrated upon the impact and effect of science-technology or technological science. It has, in effect, taken Big Science much more for granted than any of the dominant or older strands of philosophy of science.

It might be said that insofar as Big Science belongs to, or even creates, Big Technology, what many of the philosophers of technology have discerned by way of such big effects has stimulated *alarm.* Big Science-Big Technology has become a *global force* which poses effects upon global political and environmental levels. It is these phenomena which have drawn much of the attention within philosophy of technology.

Unfortunately, it has also led to extremes which are the counterparts of the narrow, propositional, and theoretically preoccupied parts within philosophy of science. Only in this case the figures drawing attention have been largely dystopian and technologically negative critics, convinced that technology is negatively affecting the essence of the human (Jonas), narrowing options to monodimensional choices (Marcuse), overwhelming nature itself (Ellul), etc. Thus, to the often sterile tone of much philosophy of science is counterpoised an alarmist dystopianism within some philosophy of technology.

Yet, just as philosophy of science in even its most narrow concerns continues to deal with issues of importance, however badly contexted, the same may be said to be the case with the extremism associated with some philosophy of technology. Big Science-Big Technology has become a global force, and it does have an effect upon the natural and social environment. And these issues must be recognized and discussed within a philosophy of technology, which is necessarily broader than philosophy of science per se—at least, so long as philosophy of science remains theory-centered. Were philosophy of science to concentrate upon *technoscience,* the outcome might bring together in a new way what now remain separate but related disciplines.

I shall point to three such areas within such a technoscience approach. The examples I select are all central to much philosophy of technology in its ethical, social-political concerns. One of those concerns relates to the *environment,* a frequently treated issue in philosophy of technology writing. My examples are: (a) PCBs and the array of chemically created compounds which are toxic and do not occur in nature but are artifacts of technoscience production; (b) nuclear energy and warfare products which imply large social-political results; and (c) industrially produced aerosols, particularly those which affect ozone in the atmosphere. In each of these cases, the product is one which does not occur within nature or within earth's environment. These products, not unlike the *produced* entities becoming paradigmatic as results of experiment, are technoscience entities. The product is created or produced through technoscience, and in each case there is, or is implied, a global effect. I am pointing to this simply as an important philosophy of science/philosophy of technology interface. The nest of issues is large, complex and urgent. But it is also an index for taking note of the areas where the latter subdiscipline does not overlap the former. There is a sense that philosophy of technology is and ought to be broader than philosophy of science. It is somewhat like the analogy between culture and religion: everyone has a culture, but not everyone is religious; everyone is involved with technology, but not all with science.

Philosophy of technology, if it is to deal with the broader issues of technology within human life, must turn its focus to issues of daily life, to the ethical impact of technologies—whether science-produced or not—and to the whole range of interfaces of technology and our lifeworld.

There is an *existential* quality to concerns which arise out of the broader areas of philosophy of technology, related to the way in which technologies have a way of "putting our bodies on the line." Concern for the natural and social environment is merely the broadest and most far-reaching of these existential implications. More regionally, but clearly relating to domains open for philosophy of technology, are those related to medicine, wherein the creation of new scarcity, life/death boundary, the artificial prolongation of "vegetative" life, and a host of other issues have given birth to a species of high medical technology ethicians. A second area of interest, only recently coming into philosophical purview, is the role of media. Here again is a technology-saturated phenomenon which in many ways parallels precisely the one discussed here concerning instrumentation. It might be said that in particular, cinema, television, and the auditory media, have enhanced and expanded (and transformed) a lifeworld in this mundane, daily sense in ways similar to the ways instrumentation changed classical science. There are embodiment, hermeneutic, and highly "carpentered" aspects to such media phenomena which could, and should, serve as interesting variants upon themes previously discussed here. This is particularly so with the differences between what is presumably the "world-exhibiting" the "world inventing" aims of similar instrumentation differently contexted.

The global, environmental, and regional medical and media regions do not exhaust what could, and should, be dealt with by a sensitive discipline in the philosophy of technology. These areas, however, point to a greater breadth, essential to philosophy of technology, than is found in philosophy of science. The task is one of "reading" the world through technology.

Because philosophy of technology is still in its infancy but is simultaneously engaged in such a broad enterprise, it is not surprising that it should remain as yet preparadigmatic. This can be an advantage, in that this new subdiscipline can learn from many sources. I have suggested that it has learned some things from the traditions of Euro-American philosophy—in part, one could say that philosophy of technology comes more fully out of that tradition than most subdisciplines have. But I also have suggested that even the older strands of Euro-American philosophy have not been as acutely aware of technologies in a more primary sense as they might have been. There have been indirect lessons from the new philosophies of science as well as the history of technology.

The other side also should be emphasized: Philosophy of science could learn a great deal from a "reading" of the world technologically, through interpreting things via their enmeshment with artifacts. I have suggested that the sense of concreteness, the sense of a certain "materiality" which arises from phenomenology in particular, may turn out to be much larger than is either expected or than is practiced today in the still separated professional and social practices of the philosophers who identify themselves with either subdiscipline. The deeper recognition of a thoroughly technologically *embodied* and *embedded* science is a first attempt at shading in this middle area.

This primer has been an initial look at the juncture of the concrete praxes of technoscience as it interfaces with the necessary emphasis upon materiality within philosophy of technology. And unlike even a little over a decade ago, a growing company has been found occupying this new territory.

Notes

I. INTRODUCTION

1. Rachel Laudan, ed., *The Nature of Technological Knowledge: Are Models of Scientific Change Relevant?* (Dordrecht: D. Reidel Publishing Co., 1984), p. 1.
2. Ibid., p. 1.
3. Mario Bunge, "Five Buds of Techno-Philosophy," *Technology in Society,* 1 (Spring 1979), p. 68.
4. Ibid., p. 1.
5. Don Ihde, *Technics and Praxis: A Philosophy of Technology* (Dordrecht: D. Reidel Publishing Co., 1979), p. xix ff.
6. Ibid., p. xxiii.
7. Thomas Kuhn, *The Structure of Scientific Revolutions* (Chicago: The University of Chicago Press, 1962) p. 1.
8. Edwin Layton, quoted by Laudan, p. 9.
9. Bunge, pp. 68–69.
10. Robert Cohen and Marx Wartofsky, in *Technics and Praxis,* p. xi.
11. One should recognize that institutional power is not at all the same thing as current intellectual power. As Steve Fuller has nicely pointed out in his "The Philosophy of Science since Kuhn: Readings on the Revolution That Has Yet To Come," *Choice,* December 1989, there has been "only [one] widely discussed book to extend the logical positivist program since 1977" (p. 595). Institutional power resides in the tenured positions often occupied for a long time after the intellectual trends have passed by the holders have ceased to be discussed
12. Laudan, p. 9.
13. Ibid., p. 10
14. Martin Heidegger, "The Question Concerning Technology," *Basic Writings* (New York: Harper and Row, Publishers, 1977), pp. 304–305.

II. THE NEW PHILOSOPHY OF SCIENCE

1. Thomas Kuhn, *The Structure of Scientific Revolutions* (Chicago: University of Chicago Press, 1962), p. 23.
2. Ibid., p. 23.
3. Ibid., p. 28.
4. Ibid., p. 42.
5. Ibid., p. 97.
6. Ibid., p. 100.
7. Edward Constant II, "Communities and Hierarchies: Structure in the Practice of Science and Technology," *The Nature of Technologized Knowledge,* p. 30.

8. Kuhn, p. 111.
9. Ibid., pp. 115–16.
10. Ibid., p. 117.
11. Ibid., p. 120.
12. Ibid., p. 120.
13. Ibid., p. 129.
14. Ibid., p. 128.
15. Ibid., p. 103.
16. Ibid., p. 111.
17. Edmund Husserl, *The Crisis of European Sciences and Transcendental Phenomenology,* trans. by David Carr (Evanston: Northwestern University Press, 1976).
18. Ibid., p. 380.
19. Ibid., p. 375.
20. Ibid., p. 25.
21. Ibid., p. 27.
22. Ibid., p. 26.
23. Ibid., p. 33.
24. Ibid., p. 35.
25. Ibid., p. 36.
26. Ibid., p. 49.
27. Ibid., pp. 49, 51.
28. Maurice Merleau-Ponty, *Phenomenology of Perception,* trans. by Colin Smith (London: Routledge and Kegan Paul, 1962), p. viii.
29. Ibid., p. xiii.
30. Merleau-Ponty, "The Primary of Perception and Its Philosophical Consequences," *The Essential Writings of Merleau-Ponty,* edited by Alden L. Fishan (New York: Harcourt, Brace and World, 1969), p. 48.
31. Merleau-Ponty, *The Phenomenology of Perception,* pp. x, xvi.
32. Ibid., p. xix.
33. Ibid., p. ix.
34. Ibid., p. xiv.
35. Ibid., p. 203.
36. Ibid., p. 203.
37. Ibid., pp. 52–53.
38. Ibid., p. 250.
39. Ibid., p. 252.
40. Ibid., p. 4.
41. Ibid., p. 251.
42. Ibid., p. 297.
43. Ibid., p. 263.
44. Ibid., p. 143.
45. Ibid., p. 143.
46. Maurice Merleau-Ponty, *The Visible and the Invisible,* trans. Alphonso Lingis (Evanston: Northwestern University Press, 1968), p. 212.
47. Michel Foucault, *The Order of Things: An Archeology of the Human Sciences* (New York: Vintage Books, 1973), p. xiv.
48. Ibid., p. xiv.
49. Ibid., p. xxiii.
50. Michel Foucault, *Discipline and Punish: The Birth of the Prison,* trans. Alan Sheridan (New York: Vintage/Random House, 1979), pp. 8–9.
51. Ibid., p. 11.
52. Foucault, *The Order of Things,* p. xv.
53. Ibid., p. 17.

54. Ibid., p. 27.
55. Ibid., p. 27.
56. Ibid., p. 18.
57. Ibid., p. 51.
58. Ibid., p. 50.
59. Ibid., p. 39.
60. Ibid., p. 20.
61. Ibid., p. 43.
62. Ibid., p. 43.
63. Ibid., cf. p. 74.
64. Ibid., p. 57.
65. Ibid., p. 57.
66. Ibid., p. 128.
67. Ibid., p. 129.
68. Ibid., p. 129.
69. Ibid., p. 130.
70. Ibid., p. 131.
71. Ibid., p. 131.
72. Ibid., p. 132.
73. Ibid., pp. 132–33.
74. Ibid., p. 133.
75. Ibid., p. 133.
76. Ibid., p. 133.
77. Ibid., p. 134.
78. Ibid., p. 268.
79. Ibid., p. 269.
80. Ibid., p. 229.
81. Ibid., p. 229.
82. Ibid., p. 268.
83. Ibid., p. 267.

III. PHILOSOPHY OF TECHNOLOGY

1. Kuhn, *The Structure of Scientific Revolutions,* p. 115.
2. Ibid., p. 59.
3. Ibid., p. 61.
4. Edmund Husserl, *Crisis,* pp. 360–61.
5. Ibid., p. 28.
6. Foucault, *The Order of Things,* p. 132.
7. Ibid., p. 133.
8. Ibid., p. 133.
9. Ibid., p. 133.
10. Martin Heidegger, *Being and Time,* trans. Edward Robinson and John Macquarrie (New York: Harper and Row, 1962), p. 95.
11. Ibid., p. 97.
12. Ibid., p. 118.
13. Ibid., p. 99.
14. Ibid., pp. 166–81.
15. Ibid., p. 101.
16. Ibid., p. 101.
17. Ibid., p. 102.
18. Ibid., p. 105.
19. Heidegger, "The Question Concerning Technology," pp. 295–96.

20. Ibid., p. 303.
21. Ibid., pp. 304–305.
22. Ibid., p. 294.
23. Ibid., p. 296.
24. See my "De-romanticizing Heidegger" (forthcoming).
25. Heidegger, "The Question Concerning Technology," p. 304.
26. Ibid., p. 296.
27. Martin Heidegger, "The Origin of the Work of Art," *Poetry, Language, Thought,* trans. Albert Hofstadter (New York: Harper/Colophon Books, 1971), p. 42.
28. J. Donald Hughes, *Ecology in Ancient Civilizations* (Albuquerque: University of New Mexico Press, 1975), p. 1.
29. Lynn White, Jr., *Medieval Technology and Social Change* (Oxford: Oxford University Press, 1962), p. 84.
30. Ibid., p. 98.
31. Ibid., p. 124.
32. Ibid., p. 125.
33. Lynn White, Jr., "Cultural Climates and Technological Advance in the Middle Ages," *Medieval Religion and Technology* (Berkeley: University of California Press, 1978), p. 218.
34. Ibid., p. 219.
35. Ibid., p. 249.
36. Ibid., p. 250.
37. Ibid., p. 228.
38. Francis Bacon, *The Works of Francis Bacon,* Vol. IV (London: Longman & Co., London, 1960), p. 47.
39. Ibid., p. 47.
40. Ibid., p. 48.
41. Although I have continued to treat Heidegger in the narrative of the text as if he were any other philosopher, no one familiar with Heidegger scholarship of the last few years can ignore the darker side of this admittedly deep thinker, i.e., Heidegger's involvement with National Socialism. Indeed, Michael Zimmerman's *Heidegger's Confrontation with Modernity,* published in this series last year, may be the most balanced treatment of that painful topic. In this context, the romanticism I accuse Heidegger of can be seen in the context of the National Socialist view to expand to the romanticism of *Volk,* the German culture, and a preference for the countryside (which, of course, while rhetorically part of the propaganda of National Socialism, was hardly the behavioral outcome of *Blitzkrieg.*) A much longer and sustained close but critical stance can be found in my own *Technology and the Lifeworld,* also in this series last year.

IV. THE EMBODIMENT OF SCIENCE IN TECHNOLOGY

1. Alfred North Whitehead, *Science and the Modern World* (New York: New American Library, 1963), p. 107.
2. Hubert Dreyfus, *What Computers Can't Do* (New York: Harper and Row, 1972), p. 147.
3. Ibid., pp. 148–49.
4. Ibid., p. 149.
5. Ibid., p. 149.
6. Ibid., p. 151.
7. Ibid., p. 153.

8. Ibid., p. 153.
9. Ibid., p. 153.
10. Ibid., p. 160.
11. Ibid., p. 162.
12. Ibid., p. 162.
13. Ibid., p. 167.
14. Ibid., p. 147.
15. Patrick Heelan, *Space Perception and the Philosophy of Science* (Berkeley: University of California Press, 1983), p. 178.
16. Ibid., p. 203.
17. Ibid., p. 142.
18. Ibid., p. 174.
19. Ibid., p. 193.
20. Ibid., p. 188.
21. Ibid., p. 198.
22. Ibid., p. 193.
23. Ibid., p. 193.
24. Ibid., p. 204.
25. Ibid., p. 203.
26. Ibid., pp. 185–87.
27. Ian Hacking, *Representing and Intervening* (Cambridge: Cambridge University Press, 1983), p. 169.
28. Ibid., pp. 158–59.
29. Ibid., p. 149.
30. Ibid., p. 167.
31. Ibid., p. 181.
32. Ibid., p. 262.
33. Ibid., p. 275.
34. Ibid., p. 208.
35. Ibid., p. 186.
36. Ibid., p. 189.
37. Ibid., p. 168.
38. Ibid., p. 191.
39. Ibid., p. 201.
40. Ibid., p. 167.
41. Ibid., p. 168.
42. Ibid., pp. 168–69.
43. Ibid., p. 193.
44. Ibid., p. 195.
45. Ibid., p. 193.
46. Ibid., p. 193.
47. Ibid., p. 194.
48. Ibid., pp. 197–98.
49. Ibid., p. 209.
50. Ibid., p. 206.
51. Ibid., p. 207.
52. Robert Ackermann, *Data, Instruments, and Theory* (Princeton: Princeton University Press, 1985), jacket.
53. Ibid., p. 8.
54. Ibid., p. 29.
55. Ibid., pp. 49, 187.
56. Ibid., p. x.
57. Ibid., p. 34.
58. Ibid., p. 73.

59. Ibid., p. xi.
60. Ibid., p. 30.
61. Ibid., p. 134.
62. Ibid., p. 32.
63. Ibid., p. 88.
64. Ibid., p. 27.
65. Ibid., p. 9.
66. Ibid., pp. 129–30.
67. Ibid., p. 13.
68. Ibid., p. 50.
69. Ibid., p. 129.
70. Ibid., p. 33.
71. Ibid., p. 63.
72. Ibid., p. 73.
73. Ibid., p. 83.
74. Ibid., p. 84.
75. Ibid., p. 87.

V. INSTRUMENTAL REALISM

1. Ian Hacking, *Representing and Intervening,* p. 176.
2. Ibid., p. 176.
3. Patrick Heelan, *Space Perception and the Philosophy of Science,* p. 177.
4. Ian Hacking, pp. 182–83.
5. Robert Ackermann, *Data, Instruments and Theory,* p. 84.
6. Ibid., p. 31.
7. Ibid., p. 9.
8. Hacking, p. 234.
9. Ibid., p. 180.
10. Ibid., p. 180.
11. Ibid., p. 180.
12. Heelan, p. 193.
13. Ibid., p. 193.

VI. THE EXPERIMENT

1. The Rachel Laudan collection, *The Nature of Technological Knowledge,* encompasses a group of post-Kuhnian revisionists. I draw attention particularly to the late Derek Desolla Price, who published and coined the term, *Little Science, Big Science* (1962) as well as accounted for much of the role of instrumentation within science.
2. Steve Fuller and Joseph Rouse, who frequently collaborate, are two such philosophers of science. Fuller's previously cited survey, "The Philosophy of Science since Kuhn," identifies many of the currently productive authors and directions. One should add here Gary Gutting, whose *Foucault's Archaeology of Scientific Reason* has just appeared (1989).
3. In addition to introducing the Heidegger/Foucault pairing to many philosophers outside much of the Euro-American circle, Dreyfus, in collaboration with his brother, Stuart, published an update on *What Computers Can't Do* in 1986—*Mind Over Machine.* I have not discussed this book which clearly updates and makes much more sophisticated the previous critique of artificial intelligence—however, the Dreyfus brothers clearly maintain the critical stance concerning newer designs.

4. Joseph Rouse, *Knowledge and Power: Toward a Political Philosophy of Science* (Ithaca: Cornell University Press, 1987), p. 23.
5. *The Scientist,* July 10, 1989, p. 11.
6. Peter Galison, *How Experiments End* (Chicago: The University of Chicago Press, 1987), p. 14.
7. Hacking, *Representing and Intervening,* pp. 22–23.
8. Galison, p. ix.
9. Ibid., pp. 2–3.
10. Ibid., p. 18.
11. Ibid., p. 14.
12. Ibid., p. 15.
13. Ihde, *Technics and Praxis,* p. 47.
14. Galison, p. 127.
15. Bruno Latour, *Science in Action* (Cambridge: Harvard University Press, 1987), p. 87.
16. Galison, p. 237.
17. Ibid., p. 238.
18. Ibid., p. 238.
19. Ibid., p. 241.
20. Latour, op. cit., p. 87.
21. Latour's Janus-faces follow closely the tradition of the "duck/rabbit" perceptual reversals followed by Kuhn and the gestalt reversals earlier analyzed by phenomenologists. But the *deliberate* use of inversions seems much closer to both phenomenological and deconstructive strategies. (See my *Experimental Phenomenology,* 1976.)
22. Latour, p. 94.
23. Ibid., p. 97.
24. Ibid., p. 99.
25. Ibid., p. 98.
26. Ibid., p. 21.
27. Ibid., p. 29.
28. Ibid., p. 259.
29. Ibid., p. 30.
30. Again, upon reading Latour for the first time, I was struck by the many parallels between his research on science and some of my own earlier work. His view of citation use and scholarly literatures applies to more than science contexts (see my analysis of the citation "power practice" in *Consequences of Phenomenology,* 1986, regarding the conflict between Anglo- and Euro-American traditions in philosophy).
31. Latour, p. 59.
32. Ibid., p. 60.
33. Ibid., p. 61.
34. Ibid., p. 69.
35. Latour's analysis of the contestative and exclusive strategies incorporated in Big Science echo and provide much ammunition to the also important school of feminist critics of contemporary science (see, especially, Sandra Harding's *The Science Question in Feminism,* 1986).
36. Latour is clearly well versed in recent French philosophy. He does cite Foucault, and there is simply too much of the annales view of history, Foucault's radically different epistemes, to disregard this as an important background to Latour's analysis. I suspect—but do not know—that the work of Jacques Derrida plays an important role here, too. The notion of "inscription" and how it occurs in the visual display is too close to dismiss.
37. Latour, p. 67.

38. Ibid., pp. 67–68.
39. Ibid., p. 65.
40. Ibid., p. 67.
41. Ibid., p. 66.
42. Ibid., p. 71.
43. Ibid., p. 71.
44. Ibid., p. 75.
45. Ibid., p. 75.
46. Ibid., p. 79.
47. Ibid., p. 79.
48. Ibid., p. 81.
49. Ibid., p. 87.
50. Ibid., p. 87.
51. Ibid., p. 87.
52. Ibid., p. 89.
53. Ibid., p. 89.
54. Ibid., p. 91.
55. Ibid., p. 91.
56. Ibid., p. 91.
57. Ibid., p. 93.
58. Ibid., p. 93.

EPILOGUE

1. Robert Crease is a colleague of Heelan's and mine at Stony Brook. He currently is the official historian of the Brookhaven National Laboratory as well as assistant professor of philosophy. In addition, I should like to point to a former student of Hacking, Davis Baird, now an assistant professor at the University of South Carolina. Baird is the author of "Five Theses on Instrumental Realism," *PSA*, Vol. 1, 1988, in which he thought he had coined the term "instrumental realism." We met for the first time in the spring of 1988 (I have used the term since 1977).
2. After the 1979–1985 period examined concerning my original list of instrumental realists, a great many philosophers of science began to turn to special interest in the experiment. This includes the subsequent projects of both Heelan and Hacking. Heelan is at work on a book, tentatively titled *After Experiment*, and has published an article, "After Experiment: Realism and Research," *American Philosophical Quarterly*, Vol. 26, No. 4, October 1989, in which he places the role of instrumental realism in the complex social context of modern experiment. Similarly, Hacking has published a recent article in which he deals with the issues of "lenses" in astronomy.
3. Especially relevant to scientific instrumentation, however, is Chapter 5, "A Phenomenology of Technics," which, while not restricted to scientific instruments, expands upon and refines much of the earlier phenomenology of instrumentation found in *Technics and Praxis*.

Index

Don Ihde is Leading Professor of Philosophy at the State University of New York at Stony Brook. He has been a principal developer of the field of philosophy of technology in North America and is the author of several books in that area, including *Technics and Praxis, Existential Technics,* and *Technology and the Lifeworld: From Garden to Earth.*